THE MEDICAL PERSPECTIVES SERIES

Advisors:

D.R. Harper *Department of Virology, Medical College of St Bartholomew's Hospital, London, U.K.*

Andrew P. Read *Department of Medical Genetics, University of Manchester, Manchester, U.K.*

Robin Winter *Institute of Child Health, London, U.K.*

Oncogenes and Tumor Suppressor Genes
Cytokines
The Human Genome
Autoimmunity
Genetic Engineering
Asthma
DNA Fingerprinting
Molecular Virology
HIV and AIDS

Forthcoming titles:

Human Vaccines and Vaccination
Antimicrobial Drug Action
Antibody Therapy

HIV and AIDS

K.E. Nye and J.M. Parkin
Department of Immunology, Medical College of St Bartholomew's Hospital, 38 Little Britain, West Smithfield, London EC1A 7BE, U.K.

First published 1994

A CIP catalogue record for this book is available from the British Library.

ISBN 1 872748 96 1

BIOS Scientific Publishers Ltd
St Thomas House, Becket Street, Oxford OX1 1SJ, UK
Tel. +44 (0)1865 726286. Fax +44 (0)1865 246823

DISTRIBUTORS

Australia and New Zealand
DA Information Services
648 Whitehorse Road, Mitcham
Victoria 3132

India
Viva Books Private Limited
4346/4C Ansari Road
New Delhi 110 002

Singapore and South East Asia
Toppan Company (S) PTE Ltd
38 Liu Fang Road, Jurong
Singapore 2262

USA and Canada
Books International Inc.
PO Box 605, Herndon, VA 22070

Typeset by Marksbury Typesetting Ltd, Midsomer Norton, Bath, UK.
Printed by Information Press Ltd, Oxford, UK.

***Front cover:** the background shows an electronmicrograph of HIV-1 budding from human CEM T cell (courtesy of Professor Robin Weiss, Institute of Cancer Research, London) and the foreground illustrates the mutating virus.*

Contents

Abbreviations

ADCC	antibody-dependent cellular cytotoxicity
AIDS	acquired immunodeficiency syndrome
ARC	AIDS-related complex
AZT	azidothymidine
BAL	broncho-alveolar lavage
BCG	Bacillus Calmette Guerin
CMV	cytomegalovirus
ConA	concanavalin A
CT	computerized tomography
CTL	cytotoxic T-lymphocyte
EBV	Epstein–Barr virus
EEG	electroencephalogram
ELISA	enzyme-linked immunosorbent assay
GM–CSF	granulocyte–macrophage colony-stimulating factor
GVHD	graft-versus-host disease
HHV	human herpesvirus
Hib	*Hemophilus influenzae* type b
HIV	human immunodeficiency virus
HLA	human leukocyte antigen
HTLV-1	human T-cell leukemia/lymphoma virus type-1
IE	immediate–early
IFN	interferon
Ig	immunoglobulin
IL	interleukin
KS	Kaposi's sarcoma
LAK	lymphokine-activated killer
LAV	lymphadenopathy-associated virus
LIP	lymphoid interstitial pneumonitis
LTR	long terminal repeat
MAC	*Mycobacterium avium* complex
MALT	mucosal-associated lymphoid tissue
MHC	major histocompatibility complex
MRI	magnetic resonance imaging
MSS	minus strong stop
NHL	non-Hodgkin's lymphoma

NK	natural killer
NSI	non-syncytium inducing
PBL	peripheral blood lymphocytes
PBMC	peripheral blood mononuclear cells
PCP	*Pneumocystis carinii* pneumonia
PCR	polymerase chain reaction
PGL	persistent generalized lymphadenopathy
PHA	phytohemagglutinin
RT	reverse transcriptase
SCID	severe combined immunodeficiency
SI	syncytium inducing
SIV	simian immunodeficiency virus
TAR	Tat-responsive region
TNF	tumor necrosis factor

Preface

It is now over 10 years since the first isolation of the human immunodeficiency virus (HIV) and the AIDS epidemic continues to grow. Both clinical and scientific research have made rapid progress; possibly more resources have been directed at this disease than any other, certainly over such a comparatively short time-scale. Early detection of the virus in blood for transfusion has allowed effective preventative measures to be implemented; the clinical effects of the first antiviral agents have been evaluated and candidate vaccines are being developed. However, there is still much to be learned about the mechanisms of pathogenesis; there is still no effective prevention or cure in sight. Education is our best hope: education of those at risk can prevent the spread of infection, while education of clinicians and carers can go a long way to alleviate the suffering of those already infected. It is our hope that this small volume will assist the latter group in the management of infected individuals in their care and may also provide facts for the educators in their quest to prevent further infection.

K.E. Nye
J.M. Parkin

Foreword

Our increasing understanding of the human immune system in recent years has been strongly fuelled by insights gained from work on AIDS. This work has highlighted the critical interaction between clinical and basic science, where events in patients ask fundamental questions about the underlying biology and where our basic models of the immune system are critically assessed on the test-bed of human disease.

However, as knowledge on AIDS has blossomed, and as more and more recondite assays are used to characterize the manifold changes in the immune system, so it has become increasingly essential to retain a perspective on the main events. It has become all too easy to fix on the latest observations or on currently fashionable concepts, which have tended to dominate thinking in this field, not least because of its high public profile. This book provides a lucid and thoughtful overview of the scientific interfaces of AIDS, maintaining a balance between perspective and focus – enabling new fragments to be fitted into the wider picture as they emerge. It should enable a wide range of people to relate the key biological concepts to their clinical counterparts and vice versa.

Recently, there has been a sense of disappointment at the limitations of available therapies and at the difficulty in sustaining progress with vaccine development. This reflects the swing of the pendulum away from over-optimistic expectations of rapid progress and towards a recognition of the difficulty of translating the speed of early insights about the nature of AIDS into effective clinical interventions. The fierce gaze of public scrutiny and the desperate needs of the millions who became infected within the first decade increased the intensity of these expectations.

The disappointment is now resolving into two contrasting streams. One has been a disenchantment with science itself, leading to some quite extraordinary views about the virus and the disease among the media and some community groups, in some instances expressing their anger and frustration against the whole scientific process and the individuals who are a part of it. The other, more consecutive response has been the recognition that it will take much longer for our existing knowledge to be translated into the therapies and vaccines that we can so readily imagine.

Wiser counsels also prevail in the process of clinical trials, with the realization that there is many a slip between a laboratory observation and the demonstration of an effect on clinical endpoints, and that changes in so-called surrogate markers may be unable truly to anticipate clinical effects. There is also a sense that we will need to understand much more of the fine detail of pathogenesis for our therapeutic and preventive response to be more effective and more focused.

In other words, we are in for the long haul, and progress will be painfully slow. It thus becomes increasingly important to retain a sense of balance about the extent and the limitations of knowledge. This book helps to define where we are, as a basis for further exploration. That exploration will necessitate partnerships between the many strands of basic and clinical science, and between scientists and the patients and communities affected. It will require a great measure of patience from us all, and a high degree of mutual understanding and respect.

Professor Anthony J. Pinching
Louis Freedman Professor of Immunology,
Medical College of St Bartholomew's Hospital,
West Smithfield, London

Chapter 1

The human immunodeficiency virus

1.1 Introduction

The human immunodeficiency virus (HIV) was recognized comparatively recently with the advent of a new syndrome, the acquired immunodeficiency syndrome (AIDS). It is estimated that 14 million people world-wide are infected with HIV, and there were 2.5 million documented AIDS cases in early 1993 [1]. The numbers continue to rise. HIV affects mainly young people, being transmitted by sexual intercourse, sharing of needles and from mother to fetus. These features illustrate why such emphasis has been placed on research into the pathogenesis and treatment of this condition.

The virus is a member of the *Retroviridae* family and the Lentivirus group. This group is responsible for infections in a broad spectrum of animal species, including visna in sheep, caprine arthritis–encephalitis of goats, and immunodeficiency in the cat, cow and primate. A lentivirus causing infectious anemia in the horse was among the first viruses to be described during the first decade of this century and is still responsible for decimation of the horse population in the Far East. Infection with HIV is manifest in immune deficiency with associated opportunist infections, lymphadenopathy and neurological disorders. Characteristic of many retroviruses is their ability to cause malignant transformation of host cells as demonstrated with the human T-cell leukemia/lymphoma virus type-1 (HTLV-1), avian myoblastosis and Rous sarcoma virus.

In the early 1980s it was apparent that certain groups of the population in the USA were displaying common signs and symptoms of a novel immunodeficiency syndrome [2,3]. Initially, all the patients were homosexual men and it was suggested that recreational drugs, recurrent sexually transmitted infections or allogeneic stimulation via white cells in semen, were the etiological factors of the immunodeficiency. When it became evident that an infectious agent was involved, the search was focused on the parvo and herpes viruses, which are known to be associated with immune deficiency. However, in 1983, Francoise Barre-Sinoussi and her colleagues at the Pasteur Institute in Paris provided evidence for the involvement of a retrovirus in the lymph node of a patient with persistent lymphadenopathy, a syndrome known to be associated with AIDS [4]. The theory that this was a variant of HTLV-1 was dispelled by Luc Montagnier, also at the Pasteur Institute, who showed that his viral isolate grew to high titer in cells bearing the CD4 surface antigen, but killed them rather than immortalizing them. Many patients with AIDS were then shown to be infected with this virus, then known as lymphadenopathy-associated virus (LAV), and it could even be demonstrated in some individuals who were apparently clinically well. Around the same time, Robert Gallo in the USA detected a retrovirus from AIDS patients which he called HTLV-III as it was clearly different from the two other human retroviruses HTLV-I and HTLV-II [5]. HTLV-III and LAV were characterized and found to be the same. The virus was subsequently named the human immunodeficiency virus (HIV) and has been isolated from people with AIDS in most parts of the world. A new strain appeared in West Africa in the mid-1980s and the two subtypes had to be classified as HIV-1 and HIV-2.

1.2 The biology of HIV-1 and HIV-2

1.2.1 Structure of HIV-1 and HIV-2 and relationship to other retroviruses

The retroviruses are RNA-containing eukaryotic viruses and are represented by certain tumor viruses as well as HIV. They contain an RNA-directed DNA polymerase known as reverse transcriptase (RT). This enzyme was discovered independently in 1970 by Howard Temin and David Baltimore. Like cellular polymerase I (Pol I) it synthesizes DNA in the 5′ to 3′ direction but, unlike Pol I, RNA and not DNA is the template, hence the name 'reverse' transcriptase. The RT of the Rous sarcoma virus, a tumor-promoting virus, is an $\alpha\beta$ dimer of 65 and 95 kDa, respectively, and that of HIV-1 is 55 and 61 kDa.

Structurally, the two subtypes HIV-1 and HIV-2 are very similar. The virus consists of a core composed of p24 or p25 capsid protein containing two identical RNA strands that are associated with the RT (DNA polymerase or Pol p55, p61) and the nucleocapsid RNA-binding proteins p7, p9 (*Figure 1.1*). Two other enzymes are also located in the core capsid, a 12 kDa protease and a 12 kDa integrase. The outer membrane of the virus, usually referred to as the envelope, is a complex structure consisting of a matrix protein p17, surrounded by a lipid bilayer around which is a glycoprotein coat containing gp120 'spikes', with a transmembraneous glycoprotein gp41. The virus expresses a number of accessory

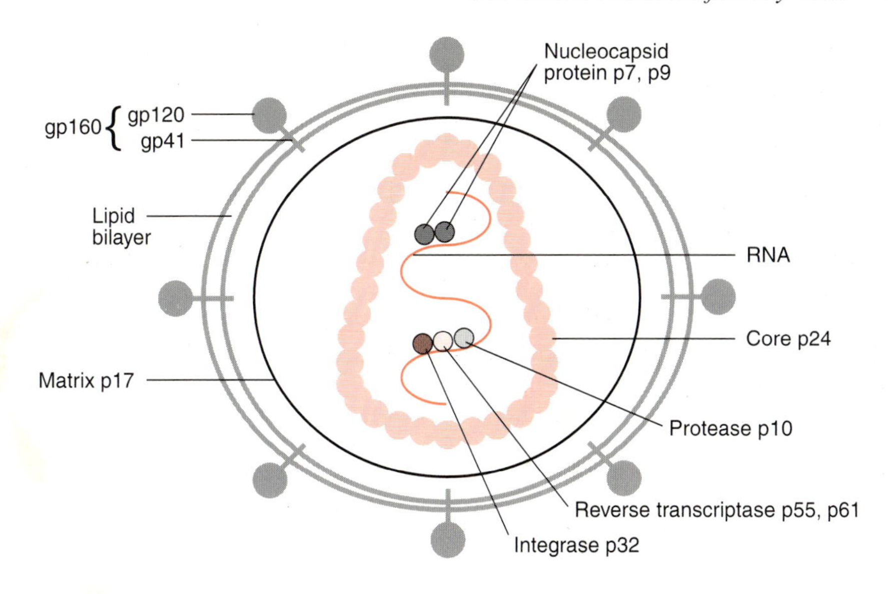

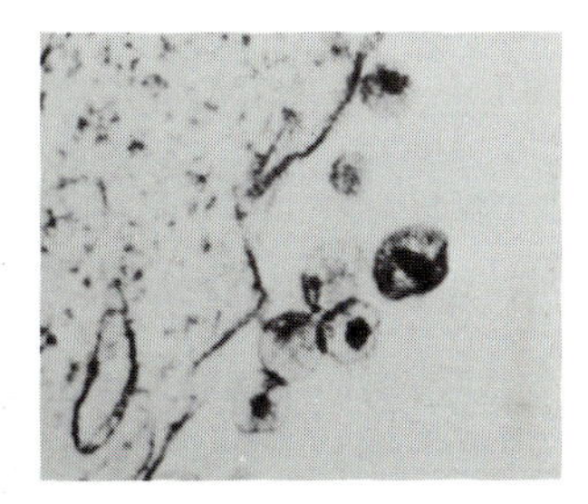

Figure 1.1: Top: structural representation of HIV. The conical nucleocapsid is surrounded by a lipid bilayer, the whole spherical structure being approximately 100 nm in diameter. Bottom: photomicrograph of HIV-1.

protein products involved in upregulation of replication or infectivity (*Figure 1.2*). The gene product associated with transactivation and the major protein involved in upregulation of replication is Tat, p14. Before viral replication can begin Tat binds to a region of the 3′ long terminal repeat (LTR) termed the Tat-responsive region (TAR) . A 23 kDa cysteine protease (Vif p23) increases the infectivity and possibly the rate of cell–cell transmission and this can be regulated by the negative factor (Nef), a myristoylated protein which is membrane associated, and also by Rev (p19), a regulator of viral mRNA expression. Other proteins expressed assist in virus release from the host cell and increase infectivity, such as Vpr (p18), Vpu (p15; found only in HIV-1), and Vpx (p15; found only in HIV-2). Tat and Rev are both activated by Tev (p26).

1.2.2 Transmission and mechanism of cellular infection

There are three major routes of viral transmission: blood; sexual contact;

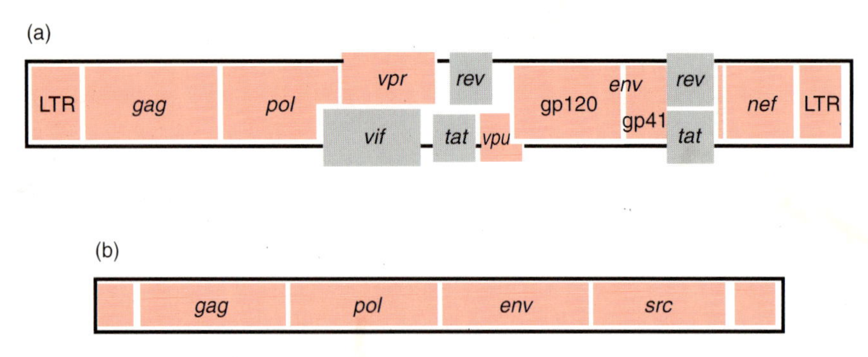

Figure 1.2: Genetic maps of (a) HIV-1 and (b) Rous sarcoma virus. HIV-1 and HIV-2 have very similar genomes. *Vpu* is only found in HIV-1 while *vpx* is exclusive to HIV-2. Both viruses encode *vpr*. The Rous sarcoma virus is typical of the RNA retroviruses. Only three of its four genes are essential for viral replication: *gag* encodes the inner core, *pol* the reverse transcriptase and *env* the outer glycoprotein. The product of the fourth gene, v-*src*, is a protein known as p60$^{\text{v-}src}$ which is responsible for host-cell transformation and is therefore termed an oncogene. HIV does not appear to encode any oncogenic gene products.

and maternal-fetal infection. Before the infectious agent was fully characterized and tested, HIV was transmitted by blood transfusion and through blood products, in particular through Factor VIII given to hemophiliacs. These routes have now been subverted, due to screening blood and organ donors for HIV, although transmission through blood still occurs in intravenous drug users. Infection associated with sexual contact was first described in homosexual and bisexual men but it was soon realized that infection also occurs efficiently through heterosexual intercourse (*Table 1.1*).

The greatest viral load is found in the blood, either as free virion in the plasma or in infected peripheral blood cells, but virus is also present in semen, breast milk, saliva, tears and cerebrospinal fluid. The number of virus-infected cells in the peripheral blood has been a matter of controversy, depending on the technique used to enumerate these cells.

Table 1.1: Transmission rates

Risk factor	Transmission rate %
Homosexual/bisexual intercourse	30–50
Heterosexual intercourse	
Male→Female	20–50
Female→Male	10–30
Maternal–fetal	10–50
Blood transfusion	Up to 100[a]
Breast milk feeding	1–5
Needle-stick injury	0.2–0.4

[a]Dependent on volume of blood and viral dose.

The most consistent figures indicate that about one in 1000 circulating mononuclear cells is infected, which means that about 5000 cells in each milliliter of whole blood contain HIV. This figure increases during symptomatic infection and in AIDS, to as much as one in ten CD4+ cells per milliliter being infected. Seminal fluid has also been shown to contain infected leukocytes, maybe as many as 10 000 infected cells in one ejaculation. Thus, these two body fluids represent the most important route of HIV transmission; tears, saliva and sweat contain negligible amounts of infectious particles. Biological factors in the infected and exposed individual also affect the chances of transmission (*Table 1.2*). Transmission is enhanced by concurrent sexually transmitted diseases, as these are often associated with genital ulceration and recruitment of inflammatory cells to the site. This leads to a loss of the physical barrier, as well as increasing the numbers of infectious cells locally in the donor and target cells in the recipient. Recipients of anal intercourse and those who have not been circumcized also appear to be at a greater risk of infection, perhaps due to tears in the rectal mucosa in the former and a greater incidence of frenal tears in the latter group during sexual intercourse.

Table 1.2: Biological co-factors for transmission

Co-factor	Infectiousness	Susceptibility
Genital ulcers	+	++
Gonorrhea	+	++
Anal intercourse		++
Lack of circumcision		(+)
↓CD4 count	++	
AIDS	++	

1.2.3 Attachment

Once the infected cell or free virion enters a new host it binds to the CD4 molecule on the surface of target cells such as lymphocytes, monocytes and antigen-presenting cells (*Figure 1.3*). Amino acids (413–447) in the conserved region of the envelope, proximal to the C-terminal of the gp120 viral glycoprotein, form the major binding site for the CD4 protein, although conformation studies indicate that other amino acid regions may also be important. A protrusion in domain V1 of the CD4 molecule is the complementary binding site for gp120 and includes a small part of the major histocompatibility complex (MHC) binding region on the same molecule. As with the gp120 molecule, it is possible that conformational changes in the CD4 molecule may expose other binding sites which could play a part in HIV entry into the cell, as well as cell–cell fusion that occurs after infection. Following binding between the HIV envelope and the CD4 antigen, it is believed that the gp120 is displaced, so exposing the fusion

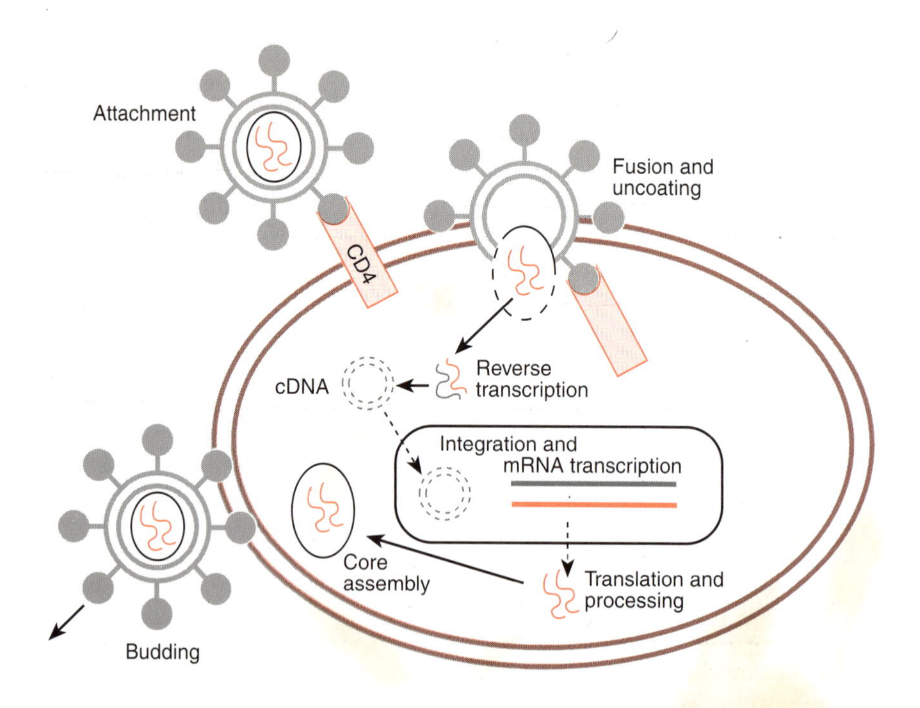

Figure 1.3: The cycle of HIV infection. Attachment occurs when a region of the gp120 viral envelope protein interacts with a domain of the host cell CD4 molecule. Conformational changes lead to fusion via the gp41 fusion domain which is followed by stripping or uncoating of the capsid to expose the viral RNA. This single-stranded RNA is converted to double-stranded DNA by the viral enzyme RT. The DNA duplex is then integrated into the host DNA and the host cell is now transformed. An event such as activation of the transformed host cell then causes transcription of the proviral DNA into viral mRNA which is further translated into precursor protein. This precursor undergoes post-translational modification leading to budding and production of a new virion.

domain of gp41, although this may vary between viral subtypes. There is some evidence for a receptor for gp41 on the CD4+ lymphocyte, although at the present time this has not been identified. HIV may also enter cells by endocytosis of the virus bound to the cell surface in clathrin-coated pits. The humoral response (B-lymphocyte-mediated antibody production) may be the initial mediator in cellular infection by HIV. Binding of non-neutralizing antibodies to the virus from either blood or semen promotes attachment to the complement or Fc receptors on susceptible host cells and can be sufficient to lead to internalization, again by endocytosis. Once within the cell, fusion of the viral envelope and vacuole membrane can take place. Whatever the means of attachment and fusion, it is at this stage that internalization and integration occurs. There is then rapid transit of these infected cells to the lymphoid tissues.

1.2.4 Fusion

The mechanism of fusion between HIV and the host cell membrane is still poorly understood. As previously stated, it was originally believed that gp120 was shed upon initial attachment. However, kinetic and thermodynamic studies argue against this hypothesis, and it is actually more likely that a conformational change exposes gp41 together with gp120 in a complex, thus presenting a different set of polypeptides to the CD4 molecule. This binding-induced conformational change may activate the fusion domain in the N-terminal region of gp41 which, together with signals transduced through the CD4 molecule, initiate the fusion step. The process appears to be pH independent, suggesting that cell surface interaction is sufficient to initiate direct cell membrane fusion, endocytic vesicles being less important.

1.2.5 Internalization and integration

Internalization of the core protein with its associated RNA and polymerases follows attachment and fusion. Once inside the host cell, the virus takes the form of a ribonucleocapsid and several events then lead to integration into the cell chromosome. Double-stranded DNA is produced from viral RNA through the process of reverse transcription. The RT of HIV-1 resides in the core and takes the form of a heterodimer, p55–p61, and is activated by an as yet unknown signal, following the binding of a host-cell t-RNA *lys* primer to an 18 nt complementary binding site near the 5′ end of the viral RNA genome. The resultant negative sense DNA is termed the minus strong stop (MSS) DNA. The viral RNase H then cleaves the original host template 5′ RNA, permitting the MSS DNA to anneal to the complementary sequence at the 3′ end of the viral RNA and thereby completing negative strand synthesis. At this stage the RNase H activity of the RT cleaves the original viral RNA from the 5′ end of the cDNA so that positive strand synthesis may begin. Integration is finally achieved when duplex cDNAs of both virus and host are cleaved by an integrase, the product of the *pol* gene.

1.2.6 Expression

Primate lentiviruses are distinguished from other retroviruses by the size and complexity of their LTRs. The regulatory elements contained within this region of HIV-1 are the polyadenylation signal site (Poly-A), the Sp-1 and NF-κB binding sites, the TATA box and TAR. There is variation in the number of copies of some of these binding sites between HIV-1 and HIV-2 (*Figure 1.4*).

The host cell regulatory element NF-κB is a most important player in viral gene expression. In the uninfected cell, this enhancer factor regulates the expression of a number of cellular genes including interleukin-2 (IL-2)

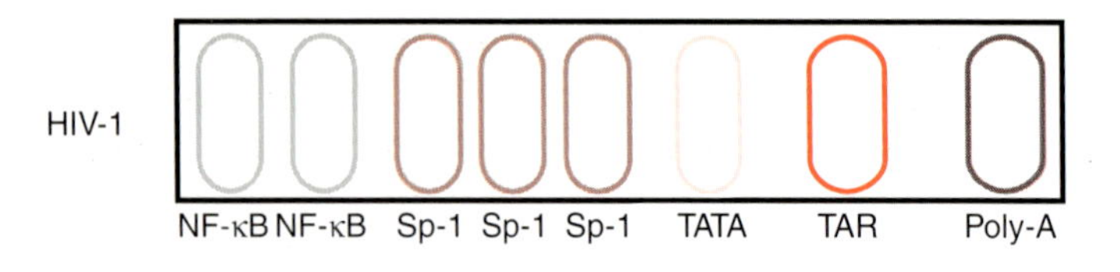

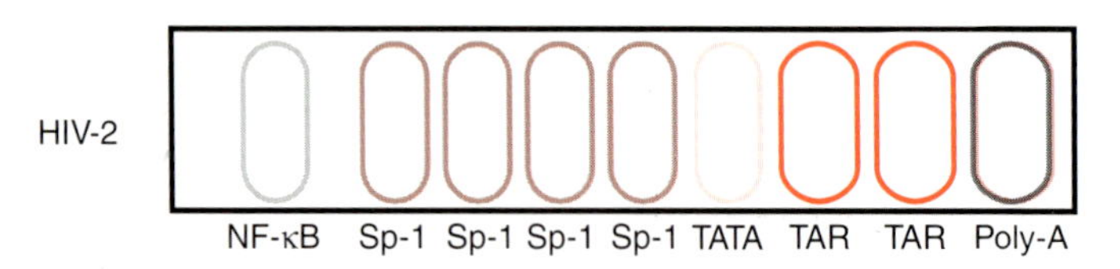

Figure 1.4: LTRs of HIV-1 and HIV-2. Both share regulatory elements such as Sp-1 binding sites, TATA boxes and the polyadenylation signal (Poly-A). The number of Sp-1 binding sites varies between the two viruses as does the overall size of the LTR (HIV-1 = 680 bp, HIV-2 = 800 bp).

and its p55 receptor, the T-cell receptor, interferon-β (IFN-β) and tumor necrosis factor α (TNFα). Activation by mitogens such as concanavalin A (ConA) and phytohemagglutinin (PHA), cytokines and viruses all induce NF-κB activity. This activation is mediated through protein kinase C or cyclic AMP, either of which cause dissociation of an inhibitor, IκB, from NF-κB which then passes into the nucleus to bind its specific enhancer elements. In HIV-infected cells activated NF-κB binds to the enhancer element present in the viral LTR and facilitates transcription of the integrated genome. It follows that cytokines that activate NF-κB in T-lymphocytes or monocytic cells will upregulate transcription of the integrated HIV genomes and may explain the role of opportunistic infections in accelerating disease progression. The transcription factor Sp-1 is of equal importance to NF-κB in the host cell mechanism and they have been shown to be interactive. Some elegant experiments to determine the relative roles of these transcription factors have been performed in which enhancer elements from the human cytomegalovirus (CMV) immediate–early gene (IE) replaced the Sp-1 and NF-κB binding sites in HIV-1 [6]. The Sp-1 and NF-κB gene products from peripheral blood mononuclear cells (PBMCs) and two T-cell lines, AA2 and CEM, were able to bind to the CMV-IE enhancer but the same factors from another T-cell line, Molt-3, or the monocytic cell lines HL60 and U937 were not. Thus, it was argued that differences in content of these transcription factors between cell types could lead to cell-specific expression of HIV-1. Subsequent experiments where LTRs were exchanged between a number of different HIV isolates suggested that either (1) the LTR alone does not dictate viral growth, tropism or pathogenicity *in vivo*, or (2) other factors such as AP-1 and NFAT-1 may also interact with the LTR. Much still has to be learned of the role of these factors in transcription of the virus.

Transcription leads to production of the regulatory elements of HIV which will control further viral replication, as well as the structural and enzymatic proteins that together with transcribed RNA will form the new virions. These proteins come together and bud through the cytoplasmic membrane, to be released into the extracellular space and infect new cells. The envelope is therefore formed partly of the host cell membrane and contains host elements. The protease enzyme is critical to the maturation of the particles, and if it is inhibited, defective, non-replicative virions are formed.

1.3 Co-factors in HIV replication

In addition to the effect of co-factors involved in the acquisition of HIV described above, it has long been thought that factors might influence disease progression by enhancing viral replication. These co-factors can mediate their effect at several levels, and may be additive. Primary infection of cells, mediated by the binding of HIV via gp120/gp160 (gp160 being a dimer of the gp120 envelope glycoprotein with the gp41 transmembrane glycoprotein) to the CD4 antigen expressed on the surface of a subset of cells, has been described above. It is likely that routes of entry other than through CD4 are also employed by the virus. It has been shown that human fibroblasts may be made susceptible to infection with HIV by CMV-induced Fc receptor expression and subsequent binding of antigen–antibody complexes [7]. There have also been reports of some strains of human herpesvirus type-6 virus (HHV-6) giving rise to CD4 antigen expression on the surface of CD8 cells.

Co-factors not only increase the absolute number of target cells permissive for infection by HIV, but have also been implicated in activating HIV gene expression. This expression, being dependent upon host-cell transcription factors, may be upregulated by pathogens that induce the production of cytokines which then activate NF-κB in T-lymphocytes or monocytic cells. Similarly, infection with opportunistic viruses, such as adenoviruses, herpesviruses, hepatitis-B virus or papovaviruses, whose early gene products are able to activate HIV gene expression through binding to unique sequences in the LTR, may also affect the pathogenicity of HIV infection.

1.4 Cellular control of HIV replication

Intracellular control of HIV replication varies between cell types in a single donor and in identical cells from different donors. Macrophages emanating from different sites vary in their infectibility. Studies show that the virus binds to each of these cells and gains entry with a similar kinetic profile; however, once inside the macrophage, events leading to integration differ from one population to another. Thus, the titer of new virus produced varies widely. Not only do the genotype and phenotype of the host cell have a bearing on viral replication, but also the viral phenotype. This is a reflection of the relative amount of host-cell-activated transcription factors present in each cell type and their interplay with the appropriate binding sites in the viral LTR. Many investigations have focused on the cellular factors that mediate functions within the viral gene through *cis*-acting elements of the viral DNA and RNA. Novel therapeutic strategies may be developed when the relationship between elements within the LTR, such as Nef, Rev and Tat, and host-cell factors, including AP-1, NFAT-1, Sp-1 and NF-κB, is found. Positive and negative regulation of viral RNA synthesis, mediated by interaction of these elements, might explain the unique pathology of HIV infection. Exposure to and entry of the virus with acute symptoms caused by primary replication is followed by a long period of latency with or without intercurrent infection, and eventually accelerated replication giving rise to the severe immune deficiency defining the diagnosis of AIDS. Mutant strains of HIV-1 and HIV-2 display altered cellular tropism, the level of

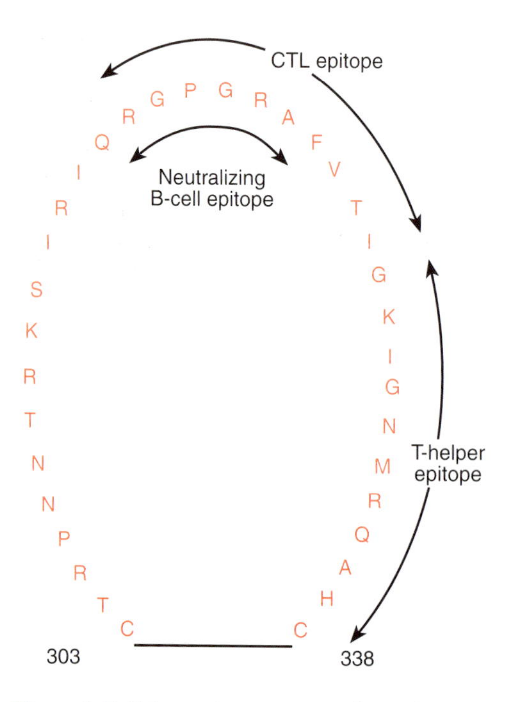

Figure 1.5. Schematic representation of V3 loop epitopes. The V3 loop is a hypervariable region within the outer surface unit of HIV-1 envelope glycoprotein. It is an immunodominant epitope and determines many properties of the virus such as tropism and infectivity. The 35 amino acids that form the loop are disulfide linked. The subdomain around the B-cell neutralizing loop, Arg-Gly-Pro-Gly-Arg-Ala-Phe, is relatively conserved in all viral isolates studied. CTL, cytotoxic T-lymphocyte. Courtesy of Linda Ebbs, British Biotechnology, Oxford, UK.

virus production varies as a function of replication kinetics and the ability to induce syncytia appears to play a vital role in disease progression. It has been shown that mutating a conserved arginine in the V3 loop (*Figure 1.5*) of the gp120 molecule alters the potential for cleavage by a cellular protease [8], which has implications for the ease of viral entry, while a further mutation at this site converts a non-syncytium-inducing virus (NSI) into a syncytium-inducing (SI) virus, with an associated rise in the replication kinetics of the virus (see p. 19).

Much has still to be learned of the molecular mechanisms of infection by primate lentiviruses and of HIV in particular. Knowledge of these mechanisms will lead to a greater understanding of the disease process in the HIV-infected individual and the subsequent progression to AIDS, and may provide the basis for novel therapies to prevent infection or halt disease progression.

References

1. WHO *The HIV/AIDS Pandemic 1993 Overview* (Document WHO/GPA/CNP/EVA/931). Global Programme on AIDS, Geneva.
2. Gottlieb, M.S., Schroff, R. and Schanker, H. (1981) *New Engl. J. Med.,* **305,** 1425–1431.
3. Friedman-Kein, A.E., Laubenstein, L.J. and Rubinstein, P. (1982) *Ann. Intern. Med.,* **96,** 693–700.
4. Barre-Sinoussi, F., Chermann, J.C. and Rey, F. (1983) *Science,* **220,** 868–870.
5. Gallo, R.C., Salahuddin, S.Z. and Popovic, M. (1984) *Science,* **224,** 500–503.
6. Chang, L.J., McNulty, E. and Martin, M. (1993) *J. Virol.,* **67,** 743–752.
7. McKeating, J.A., Griffiths, P.D. and Weiss, R.A. (1990) *Nature,* **343,** 659–691.
8. DeJong, J.J., Goudsmit, J., Keulen, W., Klaver, B., Krone, W., Tersmette, M. and DeRonde, A. (1992) *J. Virol.,* **66,** 757–765.

Chapter 2

The immunological response to HIV

2.1 Humoral immunity in HIV infection

Both arms of the immune system, humoral and cell-mediated, are invoked in response to infection by HIV. The humoral response, involving both immunoglobulin M (IgM) and IgG, is detectable within weeks of acute infection and IgG antibodies generally persist throughout the individual's life. High levels of HIV-specific IgA can be found in the gut secretions and breast milk. Antibodies are directed against the structural, regulatory and enzymatic elements of the virus. Some of the envelope-reactive antibodies have neutralizing activity; that is, they can block viral infection. The part played by neutralizing antibodies in the pathogenesis of HIV infection has been a matter of controversy in the past, largely as a result of variation of assay systems from one laboratory to another. Greater standardization of laboratory tests in recent years has produced a clearer picture of the true role of the antibody in HIV infection. However, there are still pitfalls. Neutralization assays against laboratory-adapted strains of the virus give different results for the same tests performed against autologous virus isolated from the patient or 'near patient' isolates from other individuals. Vaccine-induced antibodies usually only neutralize the virus strain which was used for vaccination, and are not cross-neutralizing with other isolates. It is unlikely that antibodies have any role in protection from cell–cell transmission of HIV (via the fusion of infected and uninfected cells), which may be the main way in which HIV is spread once infection is established *in vivo*. However, antibodies in secretions may prevent the initial infection. The stimulation of secretory antibodies by local vaccination is of great interest. The communication of B-lymphocytes between all sites of mucosa-associated lymphoid tissue (MALT) means that vaccination at one site leads to a generalized response [1].

2.2 Epitopes involved in neutralizing antibody production

Most studies show that the majority of neutralizing antibodies are directed at sites within the envelope glycoprotein. The most important domain appears to lie in the third variable region (V3 loop; see *Figure 1.5*), between amino acids 308 and 322. Although by definition the peptides in this loop vary, this short stretch shows less variation than the rest of the loop and antibodies raised to this site react with and neutralize a large number of HIV-1 strains. For most of the V3-directed antibodies the site is linear, although a number have been found that rely on the conformation of the loop. This latter group of antibodies fail to neutralize as many HIV-1 strains.

The binding site on the gp120 glycoprotein for the CD4 antigen of the host cell also generates a series of neutralizing antibodies that are generally conformation dependent. Antibodies of this type isolated from patients are also able to block the attachment of homologous HIV to the major host-cell receptor. It has been shown that monoclonal antibodies raised against the V3 loop and the CD4-binding site of gp120 act synergistically to neutralize HIV-1 [2,3].

A few antibodies have been isolated that react with epitopes in the gp41 glycoprotein. This antibody would presumably act after attachment of the virus to CD4 when gp120 is either removed or displaced. The relevance of any of these antibodies is still to be evaluated *in vivo*. Patients with AIDS still produce high titers of neutralizing antibodies when measured against laboratory strains of HIV-1, although on the few occasions that these antibodies have been tested against homotypic strains of the virus, the titer has been considerably lower. It is even possible that these antibodies enhance rather than neutralize HIV infection, by increasing Fc-receptor-mediated uptake by phagocytic cells such as macrophages.

2.3 Antibody-dependent cellular cytotoxicity

Antibodies to the envelope proteins gp120 and gp41 described above are able to induce antibody-dependent cellular cytotoxicity (ADCC). This is a process where cells bearing antigen–antibody complexes on their surface are recognized by effector natural killer (NK) cells or by other monocytic cells that have receptors for Ig on their surface. This is one of the three major mechanisms of cellular cytotoxicity of virally infected cells (ADCC, non-specific killing by NK cells and cytotoxic T-lymphocyte destruction; *Figure 2.1*). The exact cytotoxic mechanism of these cells is unclear but probably involves cytokine-mediated activation of perforins, proteins able to make holes in the plasma membrane of infected cells and thereby lead to their destruction. The clinical relevance of ADCC in HIV infection is not known. Efficiency of immune function does not appear to have a linear relationship with ADCC activity in HIV-infected individuals; in

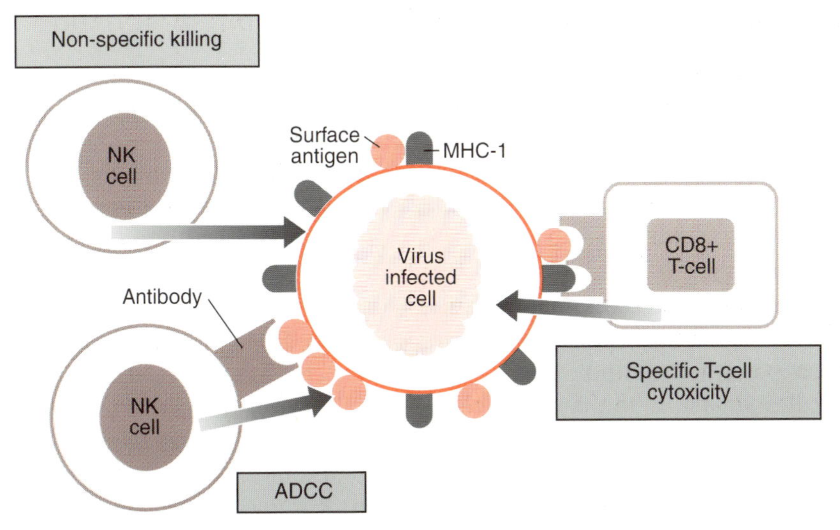

Figure 2.1: Cellular immunity in the eradication of virally infected targets. NK cells recognize virus-bearing cells by a non-specific reaction with cell surface proteins, or specifically by the interaction of cell-bound antibody with Fc receptors on the NK lymphocyte (ADCC). Cytotoxic CD8+ T-lymphocytes recognize viral antigens processed intracellularly via the Golgi apparatus and presented with class I MHC molecules at the cell surface. After recognition of the target, it is lysed by the local production of perforins which leads to loss of membrane integrity or by the induction of programmed cell death (apoptosis).

fact, some experimental evidence has shown that the release of infectious particles following ADCC-mediated destruction of the infected cell could increase infection in the host.

2.4 Cell-mediated immunity in HIV infection

There is indirect evidence to suggest that the cell-mediated immune response is the most important line of defence in the HIV-infected host. Although neutralizing antibodies persist throughout the primary stage of HIV infection, their titer only rises after the decline in viremia has occurred, and the levels of virus and p24 antigen can rise in advanced HIV infection, despite the presence of neutralizing antibody in good titer. Patients with agammaglobulinemia do not have particularly aggressive HIV disease, but if the cell-mediated system is suppressed, (e.g. by corticosteroids) then HIV progresses rapidly. Since there is a long latent stage associated with HIV-related disease, it must be assumed that immune mechanisms other than the humoral response control viral replication during this prolonged period.

One of the main protagonists to be identified in the control of HIV has been the cytotoxic T-lymphocyte (CTL). The CTL has been implicated in the immune defence against a number of viral infections, ranging from influenza to CMV and Epstein–Barr virus (EBV). Cytotoxic T-cells are generated during viral infection and are targeted at cells expressing viral antigens in association with MHC molecules (*Figure 2.2*). The MHC is divided into two classes, I and II. Antigens synthesized within the cell, particularly those from viruses, are processed via the Golgi apparatus and presented at the host-cell surface by MHC class I to T-lymphocytes expressing the CD8 antigen. Conversely, exogenous antigens are endocytosed, processed in phagolysosomes and presented by the MHC class II molecule to T-lymphocytes expressing the CD4 antigen. When the antigen presented (within the antigen-binding cleft of the MHC class I molecule) binds to the T-cell receptor on CD8+ T-lymphocytes, the infected cell begins to die. When CTLs are activated in the presence of IL-2 they proliferate and produce clones of virus-specific cytotoxic cells. The exact mechanism is not well understood but may involve either programmed cell death (apoptosis) or perforin lysis.

The CD8+ CTLs can be easily isolated directly from the peripheral blood of infected individuals. These can kill cells expressing a wide range of HIV-derived proteins, including the envelope glycoprotein (Env or gp120/gp160), core protein (p24) and reverse transcriptase. Cells responsible for killing *env*-expressing cells do not have to be MHC restricted but may be non-MHC-restricted T-lymphocytes or even non-MHC-restricted non-lymphocytes. Lymphocytes directing their cytotoxicity towards other HIV-derived proteins all appear to be classical CD3+, CD8+, MHC-class-I-restricted CTLs. These cytotoxic cells are found in abundance in the early asymptomatic period and although the numbers of CD8+ lymphocytes is relatively well maintained during disease progression, the anti-viral activity falls off rapidly. This diminished CTL activity seen in patients with AIDS is improved for a short period following therapy with the anti-retroviral drug azidothymidine (AZT). These improvements are associated with a transient restoration of a phosphatase responsible for the metabolism of the active second messengers inositol 1,4,5-trisphosphate and inositol 1,3,4,5-tetrakisphosphate [4]. This restoration of the phosphatidylinositol signaling pathway may be responsible for the improved proliferation which is also observed and could also be involved in IL-2 receptor expression and production of its cognate ligand. Loss of CTL activity has been observed in other viral diseases, including infection by EBV and CMV; this could be restored by the addition of exogenous IL-2. Hypothetically, the disturbance in the second messenger pathway could lead to loss of IL-2 and other cytokine production, leading to diminished CTL activity and apoptosis or anergy through the absence of a co-stimulatory signal. It is difficult to determine whether diminished HIV-specific CTL activity is the cause or effect of disease progression.

What part do CTLs play in the control of early infection? The experimental database in this area is small; most studies have been

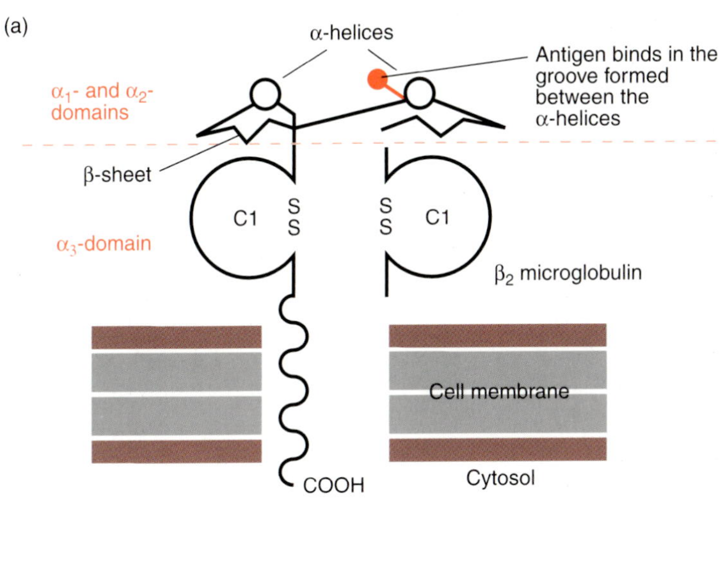

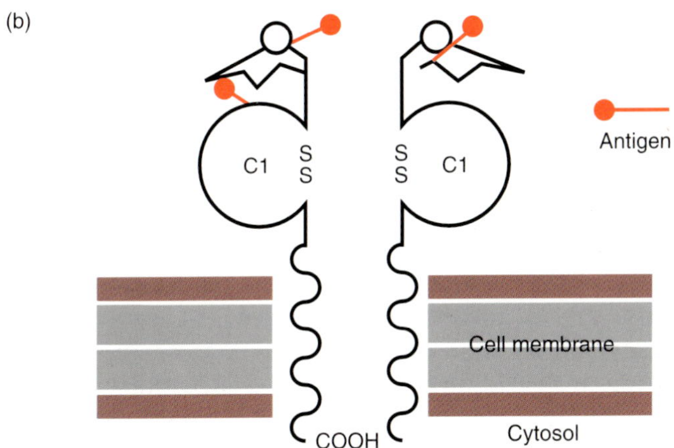

Figure 2.2: Structure of (a) class I and (b) class II major histocompatibility complex (MHC) molecules. The MHC class I molecules present endogenously synthesized peptides to CD8+ lymphocytes. The CD8 molecule interacts with the non-polymorphic α3 domain. MHC class II presents exogenously derived antigen to CD4+ lymphocytes. Recent studies have shown that the HIV and MHC class II binding sites on the CD4 molecule overlap.

performed on PBMCs from subjects with chronic infection, but what evidence exists, either from studies in rhesus monkeys [5] or humans at the time of seroconversion [6], suggests that CD8+ cells are not cytotoxic at the time of initial infection but that soon after seroconversion HIV-specific CTLs can be isolated from fresh PBMCs that are vigorously cytotoxic. Data concerning the role of CTLs in chronic infection is plentiful and the consensus opinion is that HIV-1-specific CTLs, recognizing multiple viral gene products, play a major part in attenuation of the viral load in primary infection and in maintenance of this low level during the latent phase of HIV disease. These cells may also control infection using other mechanisms as CD8+ lymphocytes from HIV-infected individuals markedly reduce viral replication *in vitro*; this suppressive effect is independent of cytotoxic activity.

There is also some indirect evidence that CTLs may confer primary immunity to HIV. The presence of HIV-specific CTLs (but not antibodies) has been shown in several groups of individuals who have been repeatedly exposed to HIV (i.e. prostitutes engaging in 'unsafe' sex with multiple partners, sexual partners of infected persons and infants born to HIV-positive mothers), but who have never become infected themselves. It is obviously ethically impossible to test the hypothesis that these individuals are now immune, but they certainly appear to have fended off HIV effectively and the part that the CTLs played is of great interest for vaccine development strategies.

On the other hand, it has also been suggested that CTLs may actually contribute to the pathogenesis of HIV disease. It has been hypothesized that CD8+ CTLs remove uninfected CD4+ cells that have surface-bound gp120 and thereby contribute to the CD4+ depletion characteristic of HIV infection. This would argue that HIV is an autoimmune disease (see Chapter 3). Recent evidence, however, shows that HIV infection is actively progressing within lymphoid tissue during the apparently latent stage of the disease and that direct HIV-mediated cytotoxicity is sufficient to explain the loss of CD4+ lymphocytes [7].

2.5 Genetic variation in HIV pathogenesis

Genetic variation may be a very important factor in HIV pathogenesis. Many genomically distinct, though related, HIV-1 strains have been isolated from one infected individual. This variation can have its effect at several different levels. The tropism for one cell type over another will affect the quality of the syndrome from one individual to another, mutant virus may escape specific humoral and cell-mediated mechanisms and virulence may be increased, affecting disease progression and severity. The substrains of HIV found in the infected individual have tropisms for different cell types. It is likely that there is a broad spectrum of genetic subtypes that will infect various cell types with equal ease and there will be those with a definite target cell. Much confusion has arisen from experiments *in vitro* where virus has been isolated from patient peripheral blood lymphocytes (PBLs) and grown in the laboratory in immortalized cell lines or activated PBLs. Few subtypes will grow in T-cell lines, while most can be propagated in the activated PBLs. Thus, some strains have been classified as slow-low or rapid-high subtypes and assigned tropism based on their performance *in vitro*, whereas the distinction may not be so hard and fast *in vivo*. HIV can also be classified as SI or NSI. The change from predominantly NSI to SI strains within the infected individual usually precedes the rapid depletion of CD4+ T-cells and the clinical diagnosis of AIDS. Whether the appearance of the SI strain causes the extreme immunosuppression or is a consequence of the progression to AIDS is difficult to ascertain.

Experimental evidence exists that shows preferential selection of mutant virus substrains that escape neutralizing antibodies and cytotoxic T-lymphocytes. This may contribute to HIV pathogenesis. The vigorous immune responses to HIV observed in primary infection that lead to the long asymptomatic period following seroconversion can be circumvented by escape mutants and propagation of these virus variants could lead to AIDS. This model is, however, too simplistic and other factors such as sheer viral load and accessory-cell dysfunction almost certainly affect disease progression.

2.6 The role of antibodies in maternal–fetal infection

Estimates of the incidence of transmission of HIV-1 between mother and child vary between 10% and 50%. Whether most of the infection that is detected in babies occurs *in utero* or at delivery is unclear. There is evidence of intrauterine infection through *in situ* hybridization and polymerase chain reaction (PCR) studies [8], and the virus has been found in cord blood and amniotic fluid. Most evidence, however, points to perinatal or post-delivery infection via amniotic or vaginal fluid and maternal blood. Infants whose cord blood was negative for HIV at birth have been found to have detectable HIV 6 weeks after delivery [9], suggesting that infection actually occurred at the time of delivery. This is reflected in studies on monozygotic twins [10] where only one twin may be infected, the first born having the higher risk factor.

Many early studies suggested a correlation between the presence of an antibody to the V3 loop of gp120 in the mother and a decreased incidence of viral transmission to the offspring. Although some recent reports do not bear this out, it is thought that this is because the experiments were performed with peptides modeled on the principal neutralizing domain of gp120, and that the presence of antibody to other parts of the V3 loop and to gp41 do correlate with protection. Other maternal factors that appear to be relevant to transmission are (1) viral load, (2) low CD4 cell count and (3) p24 antigenemia. Thus, a combination of viral strain, viral load and maternal antiviral antibody specificity might determine whether or not transmission takes place.

References

1. Lehner, T., Bergmeier, A., Panagiotidi, C., *et al.* (1992) *Science*, **258,** 1365–1369.

2. Buchbinder, A., Karwowska, S., Gorny, M.K., Burda, S.T. and Zolla-Pazner, S. (1992) *AIDS Res. Hum. Retroviruses,* **8,** 425–427.
3. Tilley, S.A., Honnen, W.J., Racho, M.E., Chou, T.C. and Pinter, A. (1992) *AIDS Res. Hum. Retroviruses,* **8,** 461–467.
4. Nye, K.E., Knox, K.A. and Pinching, A.J. (1991) *AIDS,* **5,** 413–417.
5. Reimann, K.A., Snyder, G.B., Chalifoux, L.V., Waite, B.C., Miller, M.D., Yamamoto, H., Spertini, O. and Letvin, N.L. (1991) *J. Clin. Invest.,* **88,** 1113–1120.
6. Yagi, M.J., Joesten, M.E., Wallace, J., Roboz, J.P. and Behesi, J.G. (1991) *J. Inf. Dis.,* **164,** 183–188.
7. Pantaleo, G., Graziosi, C., Demarest, J.F., Butini, L., Montroni, M., Fox, C.H., Orenstein, J.M., Kotler, D.P. and Fauci, A.S. (1993) *Nature,* **362,** 355–358.
8. Mano, H. and Chermann, J-C. (1991) *AIDS Res. Hum. Retroviruses,* **7,** 83–88.
9. DeRossi, A., Ometto, L., Mammano, F., Zanotto, C., Glaquinto, C. and Chieco-Blanchi, X. (1992) *AIDS,* **6,** 1117–1120.
10. Goedert, J.J., Duliege, A.M., Amos, C.I., Felton, S. and Biggar, R.J. (1991) *Lancet,* **338,** 1471–1475.

Chapter 3

The immunopathogenesis of HIV infection

3.1 Immunological deficiencies associated with HIV infection

It was the development of severe clinical and laboratory immunological defects in previously healthy young adults that brought AIDS to attention in 1981. There are several ways in which HIV may lead to immune defects: by direct cytopathogenicity on infected cells; by the production of immunosuppressive viral proteins; by the induction of autodestructive reactions; and by the immunosuppressive effects of associated secondary infections and tumors.

A wide range of immunological abnormalities have now been shown to be associated with HIV infection [1] (*Table 3.1*). However, the most severe defects are found within the cellular immune system, in particular the characteristic depletion of CD4+ T-lymphocytes (*Figure 3.1*). It is when the numbers of these cells in the peripheral blood drops below 200 per mm^3 that serious opportunist infections and malignancies occur. These cells function mainly as 'helper' lymphocytes, orchestrating the cell-mediated immune response by the activation of cytotoxic T-cells, macrophages and NK cells (*Figure 3.2*). They are also necessary for 'T-dependent' B-lymphocyte responses, activating cells to become antibody-secreting plasma cells, as well as downregulating IgE responses and allergic disease. This control occurs through the production of lymphokines, in particular interferon-γ (IFN-γ) and IL-2. The ability to produce these cytokines appropriately is lost even at early stages of HIV infection when CD4 cell numbers are normal, but is a bad prognostic sign for the eventual development of severe immunodeficiency and AIDS [2]. The defective cytokine production may reflect the switch from a predominant Th_1 cell population, which is characterized by the secretion of IFN-γ and IL-2, to a Th_2 population whose cytokine secretion pattern is mainly IL-4. Some investigators suggest that it is this switch that underlies the immunodeficiency.

Table 3.1: Immunological abnormalities in HIV infection and AIDS

T-lymphocytes
CD4+ lymphocyte depletion (late)
CD4+ lymphocyte dysfunction (early)
Decreased proliferation in response to soluble antigen or mitogen
Decreased IL-2 and IFN-γ production
Defective signaling via inositol pathway
Chronically elevated intracellular calcium levels
Decreased or absent delayed type hypersensitivity reactions to recall and new antigens
Defective CD8-mediated cytotoxicity
B-lymphocytes
Polyclonal activation
Hypergammaglobulinemia, IgG, IgA, IgM and IgD
Elevated levels of serum immune complexes
Auto-antibody production
Decreased ability to mount a *de novo* antibody response to a new antigen
Decreased levels of IgG2 subclass
Increased IgE-mediated allergic reactions
Other reported abnormalities
Defective antigen-presenting cell activity
Abnormal monocyte/macrophage function
Decreased NK cell function
Increased levels of acid-labile IFN-α
Anti-lymphocyte antibodies
Increased levels of β_2 microglobulin and α-1 thymosin; decreased serum thymulin levels

From ref. [1] and other sources.

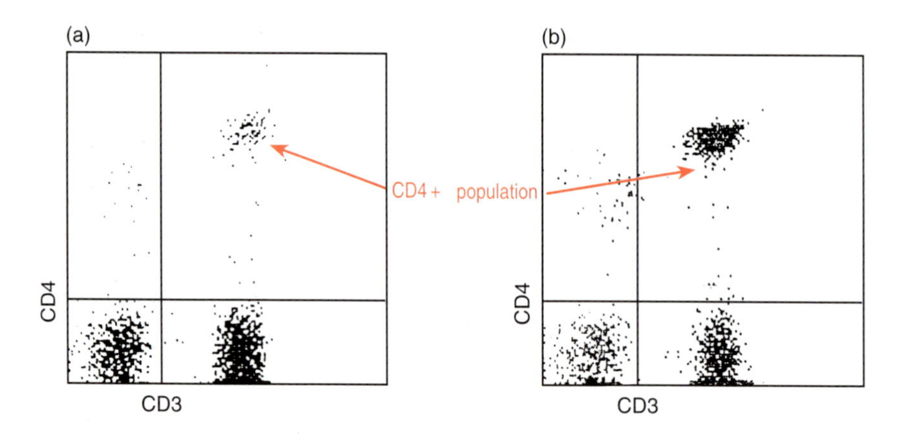

Figure 3.1: Flow cytometric analysis of peripheral blood lymphocytes from a patient with AIDS (a) and a healthy individual (b). Lymphocytes are labeled with fluorescent antibodies to the CD3 molecule (which defines all T-lymphocytes) and CD4 (which defines the CD4 subpopulation). The cell solution passes though a laser beam which activates the fluorescent markers. Each cell has the intensity of both CD3 and CD4 measured and this is demonstrated in the dot plots (each dot represents one cell). In AIDS patients, there is characteristic loss of the CD4 lymphocytes as shown.

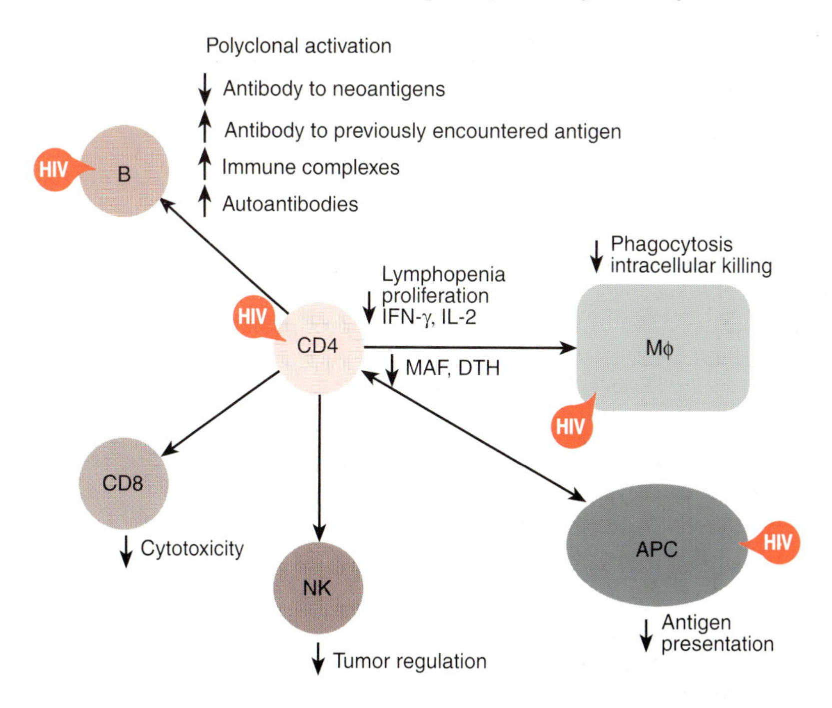

Figure 3.2: The orchestrating role of CD4 lymphocytes in the immune system and the effect of HIV on their function. CD4 cells recognize foreign antigen when it is presented with MHC class 2 antigens on antigen-presenting cells (APC) such as macrophages. Binding to the antigen receptor on the CD4 cell in this way leads to activation, proliferation and the release of lymphokines, in particular IFN-γ, IL-2, as well as macrophage-activating factors (MAF). These activate macrophages (Mϕ) to kill intracellular pathogens and produce the delayed type hypersensitivity response (DTH, via IFN-γ and MAF), enhance antigen presentation by increasing class 2 expression (IFN-γ), amplify CD8 cytotoxic lymphocyte reactions (IL-2), enhance NK cell function (IFN-γ), and regulate B-cell production of antibody (IL-2 and IL-4). HIV affects the immune response by decreasing the function and numbers of CD4 cells leading to inability to activate and regulate the other cells. In addition, HIV can infect directly APCs, including macrophages, and the viral proteins polyclonally activate B-lymphocytes.

The central role of CD4 lymphocytes means that if their function is disrupted, then it is to be expected that the function of the cells they control will also be abnormal. Therefore, the observed defects in macrophage, antigen-presenting cell and NK cell function could be due to a lack of their main activating factor, IFN-γ, and a failure to secrete IL-2 could lead to impaired cytotoxic T-lymphocyte expansion and defective B-cell or antibody responses to new antigen challenges. In addition, these cells may be directly disrupted by HIV, which infects monocytes, macrophages and other antigen-presenting cells *in vivo* [3], and can also affect B-cell and T-cell function by the production of immunomodulatory proteins, especially gp120.

In the early stages of HIV infection there is intense activation of the immune system, documented by increased numbers of CTLs bearing activation markers on their surface, the presence of immature B-cells in the peripheral blood suggesting rapid production, raised serum levels of β_2 microglobulin (a soluble form of the constituent of the MHC class I molecule which rises during lymphocyte activation), TNFα (produced by activated macrophages), and an unusual acid-labile IFN-α (which is also found in autoimmune diseases, and suggests immunological activation). This disruption in the cytokine network may have detrimental effects by inappropriately activating cells, inducing inflammation and leading potentially to apoptosis. In summary, although it is likely that the CD4 defect is the most significant deficiency in host defence, it may not explain all the features of HIV immunopathogenesis.

3.2 Mechanisms of CD4 cell depletion

Several mechanisms have been proposed to account for the severe CD4 lymphocyte depletion (*Figure 3.3*). The main debate is whether cytocidal effects of HIV on its target population are sufficient to account for the immunodeficiency, or whether other mechanisms are necessary.

3.2.1 Direct lysis of target cells

HIV isolates are highly cytopathic in cell culture, causing syncytium formation (*Figure 3.4*) and lysis by the fusion of the gp120 on the surface of infected cells with the CD4 molecules on uninfected cells. These SI variants are associated with the rapid progression of disease in humans and it has been suggested that they are the cause of immunological decline. However, using a severe combined immunodeficiency mouse (SCID) reconstituted with human immune cells to determine the pathogenicity of various HIV strains, no relation between culture characteristics and CD4 depletion has been noticed. Also, syncytia are only rarely observed in patients. Accumulation of large amounts of viral genetic material in the cytoplasm and complexing of gp120 with cytoplasmic CD4 may also be cytocidal.

The role of direct HIV infection of CD4 cells as a cause of their depletion has been questioned since only a small proportion of cells are actually infected. However, it is possible that the infected cells are destroyed so quickly they are difficult to detect in the peripheral blood. Indeed, the studies show a substantial reservoir of virus in the lymph glands of infected individuals which may explain this enigma.

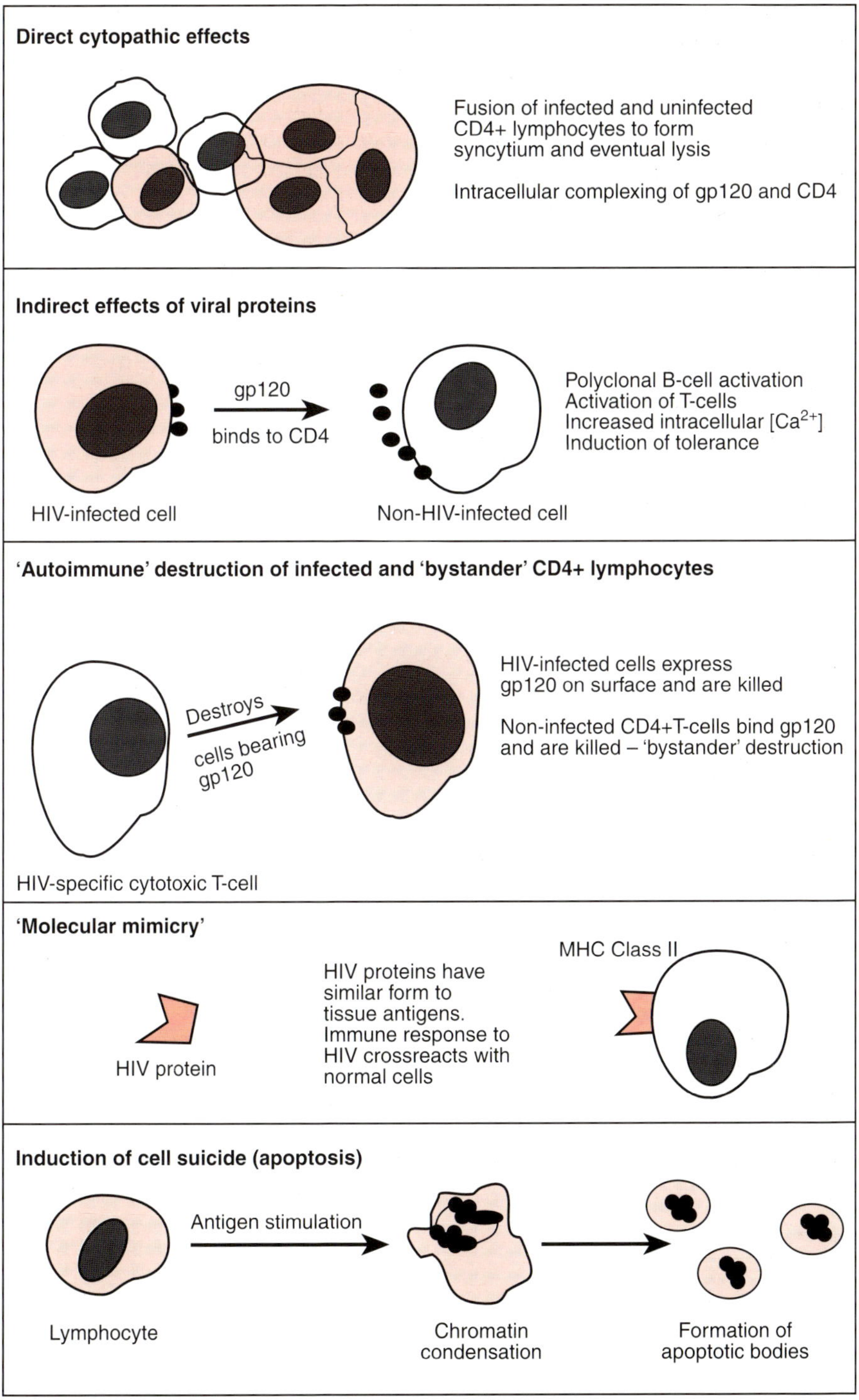

Direct cytopathic effects

Fusion of infected and uninfected CD4+ lymphocytes to form syncytium and eventual lysis

Intracellular complexing of gp120 and CD4

Indirect effects of viral proteins

gp120

binds to CD4

Polyclonal B-cell activation
Activation of T-cells
Increased intracellular [Ca^{2+}]
Induction of tolerance

HIV-infected cell

Non-HIV-infected cell

'Autoimmune' destruction of infected and 'bystander' CD4+ lymphocytes

Destroys

cells bearing gp120

HIV-infected cells express gp120 on surface and are killed

Non-infected CD4+T-cells bind gp120 and are killed – 'bystander' destruction

HIV-specific cytotoxic T-cell

'Molecular mimicry'

MHC Class II

HIV proteins have similar form to tissue antigens. Immune response to HIV crossreacts with normal cells

HIV protein

Induction of cell suicide (apoptosis)

Antigen stimulation

Lymphocyte

Chromatin condensation

Formation of apoptotic bodies

Figure 3.3: Immunopathogenesis of HIV infection.

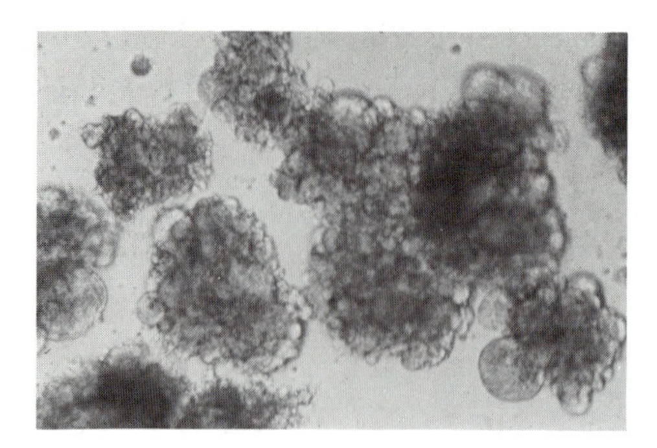

Figure 3.4: C8166 lymphoblastoid cells are highly fusogenic during HIV infection. These cells clump naturally in culture to a great degree and amplify syncytia formation. Photograph courtesy of Dr Derek Kinchington, Medical College of St Bartholomew's Hospital, London.

3.2.2 Cytokine-mediated depletion

HIV infection induces a state of immune activation, with increased production of TNFα and TNFβ, granulocyte–macrophage colony-stimulating factor (GM–CSF), IL-1, IL-3, IL-4 and IL-6. This activation may increase infection by increasing the numbers of CD4 receptors available to the virus, and by activating an infected cell to increase viral transcription.

3.2.3 'Autoimmune destruction' by HIV-specific cytotoxic cells

As described in Chapter 2, there is a vigorous cellular response to HIV infection, with the production of specific CTLs and antibodies capable of inducing an ADCC reaction. Although the removal of infected cells is desirable, it may eventually lead to depletion if they cannot be replaced. Perhaps more importantly, these cells may also recognize soluble gp120 which coats uninfected CD4+ lymphocytes, and also destroy these as 'innocent bystanders'. This mechanism is unlikely to underlie the whole story, since HIV-specific T-cells are lost with disease progression even though the CD4 depletion continues.

3.2.4 Programmed cell death or apoptosis

Apoptosis is a characteristic form of regulated cell death where there is condensation of nuclear chromatin, followed by endonuclease-induced breakage of DNA into oligonucleotide fragments, and eventually fragmentation of the cell (*Figures 3.3* and *3.5*). The particles are phagocytosed by macrophages and induce no inflammatory response. Apoptosis has been shown to be important physiologically in the removal of autoreactive immature lymphocytes during their development in the thymus. However, this response is lost during maturation and PBLs normally proliferate in response to stimulation. This clonal proliferation is of fundamental importance to the immune response in order to amplify reactions to invading organisms. However, several groups have now shown that in HIV-infected individuals, stimulation by mitogen or superantigen (an antigen that binds specific vβ chain families of the T-cell receptor and therefore stimulates a broad range of lymphocytes) leads to the death of around 40–50% of cells by apoptosis [4,5]. It affects CD4+, CD8+ and B-cells, but the CD4 population is most susceptible, particularly in late stages of the disease. This abnormal 'suicide' programming may be (1) a result of failure to generate second signals of lymphocyte activation due to intrinsic T-cell defects, (2) due to the failure of appropriate antigen presentation by accessory cells, (3) a result of the high levels of TNFα that may be produced locally, (4) due to 'lymphocyte starvation' because of a lack of production of IL-2, or (5) due to inappropriate cell signaling by soluble gp120 binding to CD4. If this phenomenon occurs *in vivo* during antigenic challenge, then clones of lymphocytes with reactivity to the particular antigen would be depleted during episodes of infection, and with time the whole repertoire may be lost.

3.2.5 Molecular mimicry

In 1986 it was hypothesized that AIDS is an autoimmune disease,

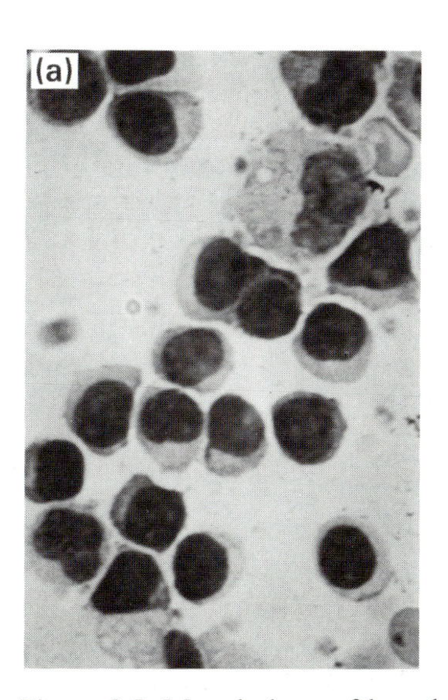

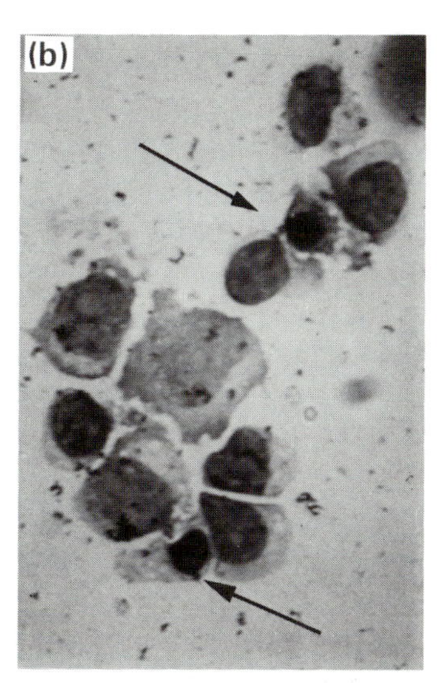

Figure 3.5. Morphology of lymphocytes undergoing programmed cell death (apoptosis), obtained from an HIV-positive individual. (a) Freshly prepared lymphocytes stained with May–Grunwald Giemsa stain. (b) Lymphocytes after 3 days culture with the mitogem phytohemagglutinin. Apoptotic cells are indicated by arrows. Photograph courtesy of Dr Nick Johnson, Medical College of St Bartholomew's Hospital, London.

triggered by HIV, and directed at the immune system [6]. It was suggested that antigenic similarity between part of the gp120 viral envelope glycoprotein and the human leukocyte antigen (HLA) class II molecule could elicit cellular immune responses or lead to the production of antiviral antibodies that crossreact with normal cells and so initiate autoimmune disease. There is evidence not only for homology between HIV and MHC genes, but also between HIV and the IL-1 receptor, IL-2 receptor and IFN genes. These observations have been further extended by Golding and co-workers [7]. They have demonstrated crossreactivity between antibodies raised against the C-terminal of HIV gp41 and parts of the HLA β-chain. Indeed, anti-class II antibodies have only been detected in HIV-positive individuals. These antibodies perturb the normal activation of CD4+ cells and may lead to the destruction of class-II-expressing cells by ADCC.

A group of biomathematicians led by Hoffmann have proposed an immune network theory that applies to the immune system in general, and have taken the findings of Ziegler and Stites and of Golding and applied them to their theorem [8]. Central to the hypothesis is the regulatory T-cell-possessing internal image idiotypes of class II MHC determinants. Radiating from this cell at the center of the network is a series of anti-self class II cells. Under normal conditions, individuals do not raise pathological antibodies against their own MHC, but with slight perturbation of the system antibodies could easily be produced against the internal image of class II cells and these reactions may be further amplified through stimulation by allogeneic lymphocytes acquired from the blood and seminal fluids from sexual contact. Hoffmann and co-workers went on to perform a series of experiments in which they immunized mice with PBMCs from allogeneic mice that subsequently produced antibodies that both bound to normal human lymphocytes and neutralized HIV. The response was most likely raised to shared MHC regions on the cell surface.

Another source of mimicry involves amino acid homologies between the CD4 molecule and microbial proteins derived from opportunistic infections associated with AIDS. Root-Bernstein and Hobbs have hypothesized that antibodies to these microbial epitopes can act as cytotoxic anti-CD4 antibodies and induce lymphocytotoxic autoimmunity [9]. More data are needed to verify this hypothesis.

The symptoms associated with AIDS have been compared to those seen in graft-versus-host disease (GVHD) and, indeed, the concept of autoreactivity mediated by retroviral infection was proposed as long ago as 1976 [10]. In this model the immune response of the HIV-infected individual is inappropriate and leads to the destruction of the host's T-cells and in particular of the CD4+ subset. Antibodies to collagen degraded by inflammatory agents are typical in autoimmune disease. These same antibodies have also been detected in GVHD and in patients with AIDS, suggesting a link between autoimmunity, GVHD and the immunopathogenesis of AIDS.

Antibodies raised against non-structural polypeptides within HIV may also lead to autoimmune responses. Sequence identities between regions of the MHC class II protein and the *nef* gene product of HIV have been reported [11,12]. Antibodies directed against the Nef protein could crossreact with the MHC structure on the lymphoid cell surface and play a role in the autoimmune pathologies associated with AIDS.

In conclusion, molecular mimicry of normal cellular proteins by viral proteins may cause an autoimmune response against the cellular components, although the role of autoimmunity in HIV-induced disease has still to be established.

3.2.6 Anti-lymphocyte antibodies

Several groups have identified anti-lymphocyte antibodies in HIV-infected individuals. Some have shown association with progression of disease. An antibody of particular interest was identified by Stricker and colleagues that reacts with an 18 kDa antigen, possibly histone H2B, which is only present on activated CD4+ lymphocytes [13]. The antibody causes lysis of cells in the presence of complement, suggesting a possible pathogenic role. Anti-CD4 antibodies have also been identified.

3.3 Mechanisms of CD4 lymphocyte dysfunction

Long before CD4 cell numbers drop, there is evidence of abnormal function and this progresses as the number of cells decline; that is, even cells that remain are ineffectual. Several reasons have been suggested for this deficiency, mainly implicating the effects of HIV proteins on uninfected CD4+ cells and antigen-presenting cells.

3.3.1 Immunosuppression by viral proteins

The viral envelope glycoproteins gp120 and gp41 have immunosuppressive effects on the mitogenic responses of human T-lymphocytes and they also perturb the activity of NK cells. This effect may be mediated through interference with one of the second messenger systems. Binding of antigens or mitogens to the T-cell-receptor–CD3 complex and the CD28 cell surface molecule affects gene expression within the cell. This is achieved by a series of protein phosphorylation and dephosphorylation events which result in the release of certain intracellular proteins from their inhibitors (e.g. NF-κB from IκB; the active NF-κB then travels into the nucleus where it binds to other proteins and, in the case of the HIV-infected cell, it binds to the HIV-LTR and can upregulate viral replication). It appears that in the case of gp120 or gp160 binding to the CD4 molecule the signaling pathway is perturbed in such a way that the cell becomes chronically activated and is refractory to further stimulation. The *src*-family-related tyrosine kinase, $p56^{lck}$, is displaced from the cytoplasmic tail of the CD4 molecule and as a consequence does not phosphorylate a number of its intracellular targets. This effect of gp120 or gp160 binding to intracellular second messengers could interfere with normal T-cell help to B-cells, either through inappropriate production of cytokines or blockade of contact-dependent interactions.

Another route to immunosuppression through viral proteins is mediated through the formation of antibody–viral antigen complexes. These can act by binding to cells of the reticuloendothelial system and thus affecting cytokine production (e.g. IL-1) and therefore immune function. How much these immunosuppressive agents affect the pathology of HIV infection *in vivo* is uncertain; it appears that the major factor is still likely to be loss of the CD4 population of T-lymphocytes.

3.3.2 Accessory cell dysfunction

The immune system is a network of cells, T- and B-cells, macrophages, dendritic cells and a host of other players, controlled either through intimate contact or by a series of chemical messengers, interleukins, interferons and other cytokines, that work in harmony to maintain the delicate balance of the healthy organism. The macrophage is a most important member of the immune series that may act as a reservoir and carrier of HIV to other cells and organs. These cells are non-proliferating and are therefore able to sustain HIV production for long periods without being destroyed by the virus. There is also evidence that cytokine secretion by the infected macrophage is aberrant [14], which may have significance for the wasting syndrome (slim disease) and central nervous system disease seen in many AIDS patients. While the network between antigen-presenting cells and effector T-cells may play a part in these diseases, it may well be that these interactions are a direct effect of the inappropriate cytokine production by the infected macrophage.

The role of the macrophage as an antigen-presenting cell, especially to the peripheral memory T-cells, does not appear to be compromised during asymptomatic HIV infection. The story may well be different in the lymph nodes. It has been postulated that both the dendritic cell and macrophage are affected in their antigen-presenting function to naive T-cells in the lymph nodes in early HIV infection [15]. If there is dysfunction in both the cytokine and antigen-presenting function of the macrophage then the cell would be a reservoir of HIV, and as an effector cell would induce anergy in the Th_1 cell population and increased activation of the Th_2 population (see Section 3.1). Memory T-cell responses may be sufficient to keep HIV-infected individuals free from secondary infection in the early stages of the disease, but since this subset is depleted, and the interaction between antigen-presenting cells and naive T-cells no longer takes place, the immune deficiency increases.

3.4 B-lymphocyte dysfunction

B-cells in HIV-infected patients are of interest as, despite never having been shown to be infected by HIV *in vivo*, their function is highly abnormal. The characteristic finding is of polyclonal activation causing raised serum levels of IgG and IgA in adults (*Figure 3.6*) and IgM in infants and increased spontaneous Ig production by B-cells *in vitro*. A significant amount of the Ig is against HIV, previously encountered infectious agents and autoantigens, (i.e. 'old immunizations'). However, these overactivated cells are unable to respond to new antigens, and are functionally hypogammaglobulinemic. This means that there is a markedly reduced response to vaccination, which is affected even at early stages of HIV infection. In addition, the hypergammaglobulinemia is mainly of the IgG1 and IgG3 subclass, and many patients have an absolute deficiency of IgG2. These individuals suffer recurrent pyogenic infections, as it is this antibody subclass that offers protection against the polysaccharide coat of encapsulated organisms (*Figure 3.7*). In adults the poor antibody response is only of minor relevance as the patients have already built up an antibody repertoire to many infectious agents during their lives. However, in children infected *in utero*, who have no such repertoire, the effect is more pronounced, and bacterial sepsis due to ordinary organisms may be as common a cause of death as *Pneumocystis carinii* pneumonia (PCP).

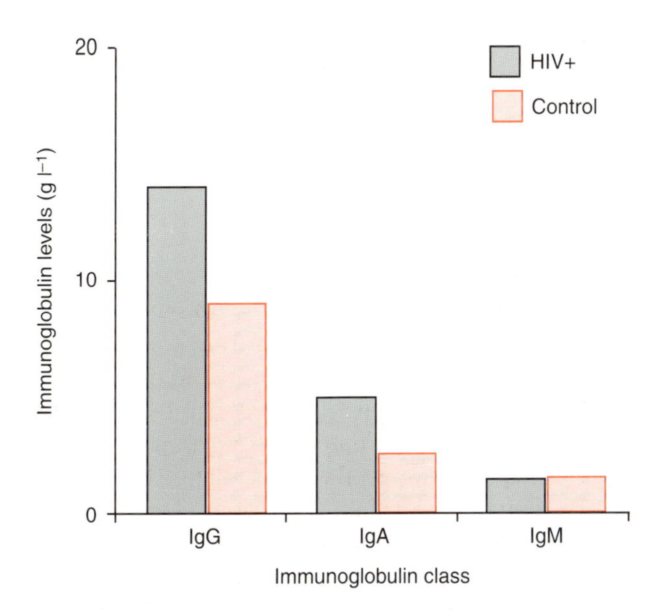

Figure 3.6. Hypergammaglobulinemia is characteristic of HIV infection. This involves predominantly the IgG and IgA isotypes in adults. This is a result of polyclonal B-cell activation and the antibody produced is mainly to previously encountered antigens.

The cause of the B-cell defect may be contact of these cells with the soluble viral proteins produced by infected cells. The envelope proteins have been shown to activate B-lymphocytes polyclonally in culture. Loss of T-cell control and failure to regulate B-cell-activating viruses, in particular EBV, may also play a role. The progressive switch from a Th_1 to Th_2 cell type during progression of HIV infection may explain the re-emergence of allergic disease which is noted in patients as they become immunosuppressed. It has been shown that the lymphokines produced by Th_1 cells, in particular IFN-γ, downregulate IgE production by B-cells, whereas those produced by Th_2 CD4+ cells, IL-4 and IL-5, increase IgE secretion.

3.5 Natural history of HIV infection

The natural history of HIV-1 is well documented, particularly from longitudinal studies of cohorts of homosexual men and hemophiliacs infected in the early 1980s [16]. Despite the fact that HIV was only discovered in 1984, some studies have follow-up times of over 14 years from seroconversion, using data obtained retrospectively from stored sera. Less is known about HIV-2, as it has only relatively recently been recognized [17], and most infected individuals are in West African countries (Guinea Bissau, Senegal and Ivory Coast), where there are less resources available for intensive investigation. The spectrum of disease appears similar to that caused by HIV-1, with lymphadenopathy and immunodeficiency [18]. However, the data that are currently available suggest that HIV-2 is less pathogenic [19]. Dual infection occurs. Another variant termed 'strain O' has been demonstrated in individuals from Cameroon; very little is currently known about this virus and it remains rare outside of Africa.

3.6 Acute infection and seroconversion

Acute HIV infection is associated with active viral replication, demonstrated by the presence of p24 antigen in the serum. Occasionally, this initial viral replication is extensive enough to cause transient CD4 lymphocyte depletion, and individuals may develop opportunist infections, including esophageal candidiasis and PCP. However, unlike patients at later stages of disease, the immune system recovers as the viral replication comes under control, and they may return to an asymptomatic phase for some time. Development of specific antibodies usually occurs within 6–12 weeks of exposure and is associated with a rapid decline in viremia (*Figure 3.8*). It is during this time that a 'seroconversion illness' occurs in approximately 10% of infected individuals. The features are non-specific and include a sore throat, generalized lymphadenopathy, arthralgia (joint pains), fevers and skin

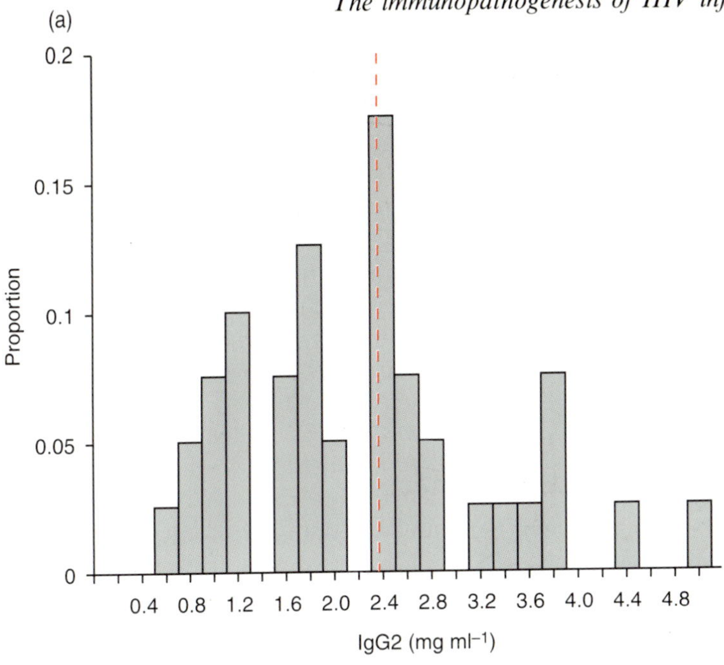

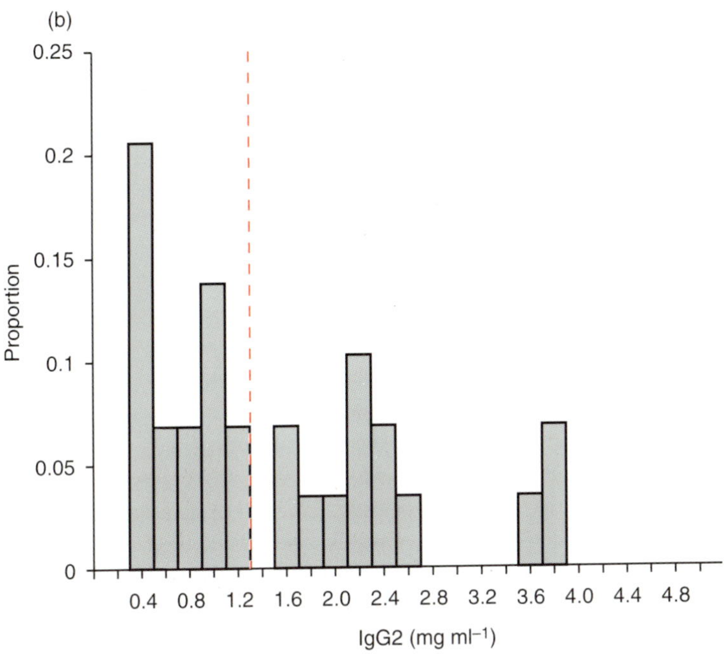

Figure 3.7. Dysgammaglobulinemia in HIV infection. The polyclonal B-cell activation leads to increased levels of IgG1 and IgG3 subclasses of antibody. Conversely, IgG2 levels may be deficient and lead to a 'functional hypogammaglobulinaemia' with susceptibility to pyogenic infection with encapsulated bacteria such as *Hemophillus influenzae, Streptococcus pneumoniae* and *Moraxella cattarrhalis*. The graphs show (a) normal levels in HIV-infected patients without pyogenic infections and (b) the low levels found in those patients with susceptibility to bacterial infections.

rash. It may be mediated in part by immune complexes. The immune response is vigorous, with up to 1% PBLs making a specific antibody to the structural, regulatory and enzymatic elements of HIV. Although neutralizing antibodies are produced, their importance to the initial control of HIV infection is uncertain, as HIV p24 antigen and the high titers of infectious virus in blood are lost before neutralizing antibody can be detected (see Chapter 2). In addition to the humoral response, CD8+ cytotoxic T-lymphocytes are found, which recognize HIV-infected targets. CD8 lymphocytes from HIV-infected individuals are known to reduce HIV replication *in vitro*, and it is likely that HIV-specific CTLs would contribute to the reduction in the peripheral blood HIV burden during these early stages. The initial immune response to HIV appears to be crucial in determining the future course of the disease. Those who make only low levels of antibody, in particular anti-p24 (anti-core) are much more likely to have a rapid progression to immunodeficiency than those who make high titer responses.

3.7 Stages of chronic HIV infection

Following seroconversion there is a variable period of clinical latency in which the individual is asymptomatic, p24 antigen is undetectable and the viral burden is low in the peripheral blood (only one in 10 000 to one in

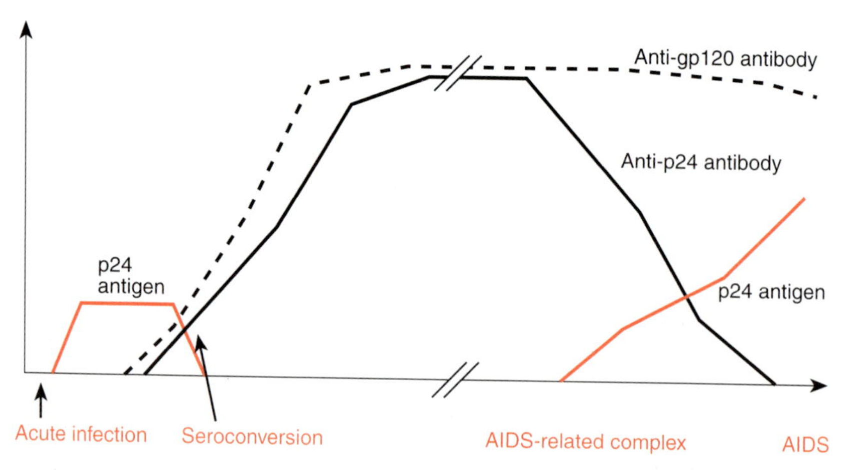

Figure 3.8. Serological features of different stages of HIV infection. HIV core antigen (p24) is the first detectable sign of HIV infection in the serum. However, its presence is only transient and is not found in all individuals, making it an unreliable test for early HIV infection. Within 4–6 weeks antibody to core (anti-p24 antibody) and envelope (anti-gp120 antibody) are consistently detectable. With disease progression p24 antibody may be lost coincidental with the re-emergence of p24 antigenemia. This is a bad prognostic sign for the development of full-blown AIDS. Other anti-viral antibodies, including that to gp120, are generally maintained.

50 000 lymphocytes containing HIV genomes). Recently, PCR has been used to give a more detailed quantitation of plasma viremia, by measuring HIV RNA and DNA in plasma (*Figure 3.9*). This shows that even at an early stage there is active viral replication. Therefore, it is clear that HIV is never truly latent.

Some develop a syndrome of persistent generalized lymphadenopathy (PGL), in which there is painless symmetrical swelling of the lymph glands of the neck, axilla and groin. This may also involve the adenoids. Analysis of the lymph nodes has given useful information about the pathogenesis of HIV, showing that there is significant HIV infection of CD4+ lymphocytes and follicular dendritic cells, even when the levels of virus in the peripheral blood are low. Histologically, there is follicular hyperplasia, with an influx of CD8+ cells (CTL phenotype) into the germinal centers. It is assumed that these cells are HIV-specific cytotoxic lymphocytes involved

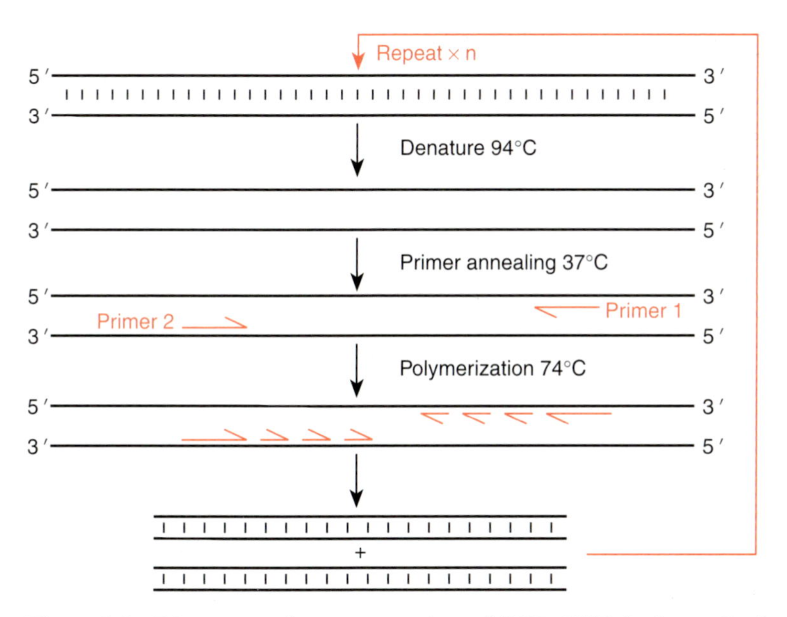

Figure 3.9. Diagrammatic representation of PCR. PCR is the cyclical amplification of a specific DNA sequence by a theromostable DNA polymerase. There are three stages within each cycle, each at a different temperature. These stages are denaturation of double-stranded DNA, which takes place at elevated temperatures to ensure complete separation of the individual strands. These form the templates for the annealing of short complimentary oligonucleotides (primers) which bind at lower temperatures. The final step is polymerization by the polymerase from the (3′) terminal of the primer as the reaction temperature is raised to 74°C. This series of steps is repeated for up to 40 cycles to achieve million-fold amplification of the sequence of DNA located between the primers. Diagram courtesy of Dr Nick Johnson, Medical College of St Bartholomew's Hospital, London.

in the control of HIV replication. Therefore, PGL may be a sign of a vigorous response to HIV. This fits in with clinical studies which have shown that individuals with PGL have a better prognosis than those without.

3.8 Progression to symptomatic disease

With time, there is progressive loss of lymph node architecture, with involution of the germinal centers, and destruction of the fibroblastic supporting structure occurs. This may be a result of the cytopathic effect of HIV, of released viral proteins or cytokines, or it may be secondary to the inflammatory actions of infiltrating cytotoxic T-cells. During the stage of lymph node involution or loss of lymphadenopathy as observed by clinical examination, there is characteristically a progressive decline in CD4 counts, and development of HIV-related disease. During the same time the viral load in the blood and the percentage of infected cells increases steadily, possibly due to the loss of localized control of the virus at lymph node sites. Increased viral replication leads to further loss of immunological control and a vicious cycle develops (*Figure 3.10*). Serology frequently shows loss of p24 antibody (despite other anti-HIV antibody levels being maintained) and re-emergence of p24 antigen in the blood. These features are bad prognostic indicators for the rapid development of AIDS.

The initial information gained on progression rates of the disease made grim reading, since there appeared to be a fixed rate of progression to AIDS of between 2 and 5% per year, leading to the conclusion that eventually all HIV-infected individuals would become sick. However, it is now apparent that after 10 years or so of follow-up, the progression rate in cohorts may slow, and around 15% of the infected individuals are remaining asymptomatic, the so-called 'long-term survivors'. There is currently intense investigation of these individuals for factors, host, environmental or viral, that could account for their prolonged survival. However, a prediction of their ultimate prognosis remains guarded, as some show evidence of immunological damage with decreasing CD4 counts despite being clinically well.

At the other end of the spectrum, there are individuals who develop full-blown AIDS within 18 months of HIV infection, the 'rapid progressors'. It has been suggested that infective and other co-factors may be involved (*Table 3.2*). Candidate organisms are HHV-6 (which can induce the expression of CD4 on CD8+ lymphocytes and make them susceptible to HIV), or organisms such as herpes virus types 1 and 2, CMV, HTLV-1 or mycoplasma that can activate HIV replication and may therefore cause rapid progression of disease. Activation of lymphocytes also activates HIV replication within them. This may occur during allogeneic stimulation during pregnancy (the fetus being partly of foreign HLA type). In practice there have been conflicting reports of the effects of pregnancy

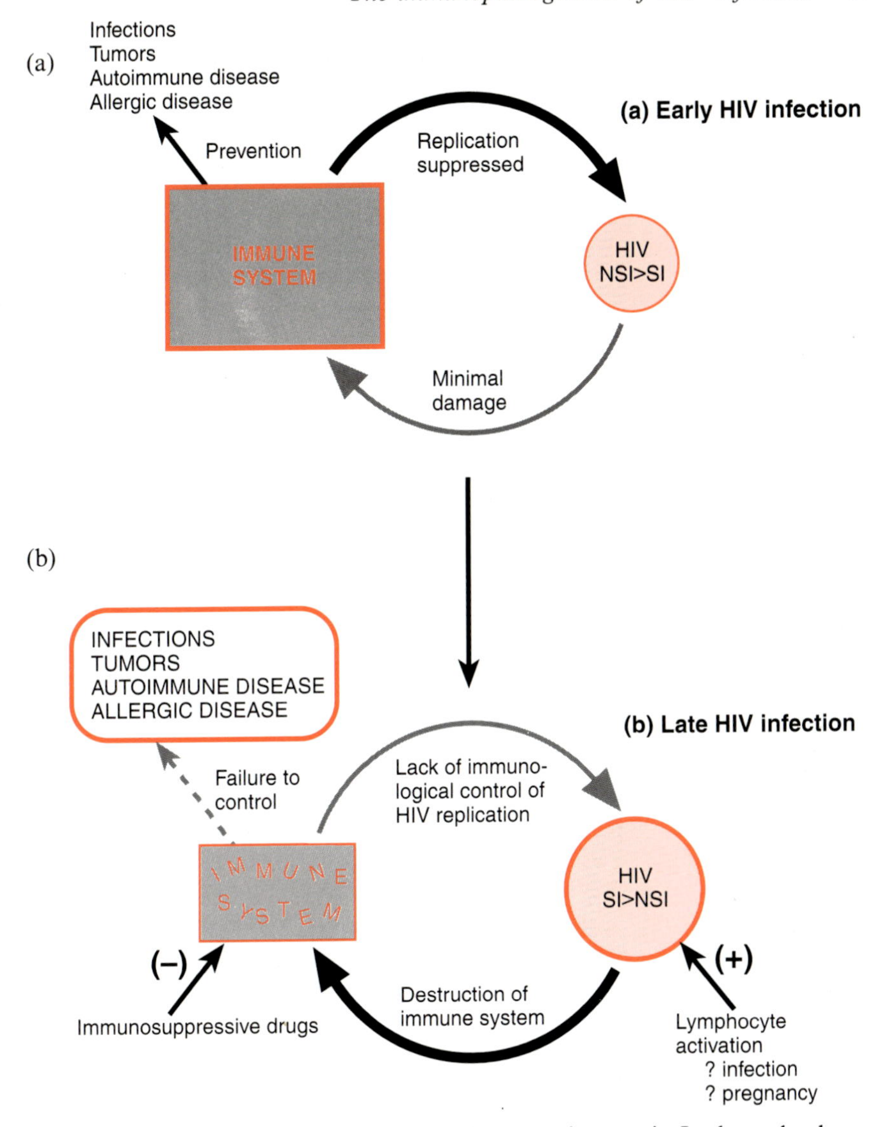

Figure 3.10. The vicious cycle hypothesis of HIV pathogenesis. In the early phases of HIV infection (a), the immune system mounts a specific response to the virus and keeps replication controlled. This prevents major immunological damage from the direct cytopathic effects of HIV, and the immune system can perform its usual role of protection against infections and tumors. The main type of virus isolated at this stage is 'non-syncytium inducing' (NSI). With time, there is increased HIV replication (b), possibly as a result of viral mutation leading to immune escape (possibly enhanced by lymphocyte activation by intercurrent events such as infections and pregnancy). The virus is now of the more pathogenic 'syncytium inducing' (SI) phenotype and there is HIV-mediated damage to the cellular immune system. Immunosuppression reduces the ability of the immune system to control viral replication, which goes ahead unchecked leading to further destruction of immunity. Immunosuppressive drugs, such as corticosteroids, used at this stage may lead to rapid disease progression, by further depressing the cell-mediated response. With disintegration of the immune system comes failure to control infections and tumors and development of opportunist disease.

Table 3.2: Factors associated with rapidly progressive HIV disease

Population	Factor
Perinatally infected infants	Advanced disease or p24 antigenemia in mother during pregnancy
Adults	Poor initial immunological response to HIV, e.g. low p24 antibody levels
	Increasing age
	Immunosuppressive agents, e.g. corticosteroids
	Lymphocyte activation
	Pregnancy?
	Opportunist infection?
	Recreational drugs??
	Virus type
	Infectious co-factors including HHV-6, herpes simplex virus, HTLV-1, CMV, mycoplasma?

on HIV-related disease. Studies from developed countries, mainly involving drug users who inject, have shown no adverse effect. However, in African studies, pregnancy does appear to enhance progression. The explanation that has been given is that drug users tend to use less during pregnancy, have better nutrition and medical care, and therefore their physical health may be maintained and compensate for any adverse effect of HIV activation.

The only factor which has consistently been shown to correlate with increased progression rates across all adult populations is age. One study has shown that the median 'treatment-free' incubation time is 12 years for HIV infection contracted at the age of 20 years, 10 years for infection at 30 years, and 8 years for infection at age 40 years [20]. Immunosuppressive drugs such as corticosteroids and cytotoxic agents may also enhance progression by impairing the immunological suppression of HIV, allowing the virus to break free. In children infected *in utero* or perinatally, the stage of disease of the mother and levels of her viremia not only determine whether the children acquire HIV infection in the first place, but also the speed at which they develop HIV-related disease [21].

References

1. Seligmann, M., Chess, L. and Fahey, J.L. (1985) *New Engl. J. Med.*, **311,** 1276–1292.
2. Murray, H.W., Rubin, B.Y., Masur, H. and Roberts, R.B. (1984) *New Engl. J. Med.*, **310,** 883–889.
3. Cheng-Mayer, C. (1990) *AIDS*, **4** (Suppl. 1), 49–56.
4. Groux, H., Torpie, R.G., Monte, D., Mouton, Y., Capron, A. and Amiesen, J.C. (1992) *J. Exp. Med.*, **175,** 331–340.
5. Meyaard, L., Otto, S.A., Jonker, R.R., Mijnster, M.J., Keet, R.P.M. and Miedema, F. (1992) *Science*, **257,** 217–219.
6. Ziegler, J.L. and Stites, D.P. (1986) *Clin. Immunol. Immunopath.*, **41,** 305–313.
7. Golding, H., Shearer, G.M., Hillman, K., Lucas, P., Manischewitz, J., Zajac,

R.A., Clerici, M., Gress, R.E., Boswell, R.N. and Golding, B. (1989) *J. Clin. Invest.,* **83,** 1430–1435.

8. Hoffman, G.W., Grant, M.D. and Kion, T.A. (1991) *Proc. Natl Acad. Sci. USA,* **88,** 3060–3064.
9. Root-Bernstein, R.S. and Hobbs, S.H. (1994) *Med. Hypotheses,* (in press).
10. Proffitt, M.R., Hirsch, M.S. and Black, P.H. (1977) in *Autoimmunity: Genetic, Immunologic, Virologic and Clinical Aspects.* (N. Talal, ed.). Academic Press, New York, pp. 385–401.
11. Morrow, W.J.W., Isenberg, D.A., Sobol, R.E., Stricker, R.B. and Kieber-Emmons, T. (1991) *Clin. Immunol. Immunopathol.,* **58,** 163–180.
12. Bjork, R.L. (1991) *Immunol. Lett.,* **28,** 91–95.
13. Stricker, R.B., McHugh, T.M., Moody, D.J., Morrow, W.J.W., Stites, D.P., Shuman, M.A. and Levy, J.A. (1987) *Nature,* **327,** 710–713.
14. Fauci, A.S., Schnittman, S.M., Poli, G., Koenig, S. and Pantaleo, G. (1991) *Ann. Int. Med.,* **114,** 678.
15. Macatonia, S.E., Lau, R., Patterson, S., Pinching, A.J. and Knight, S. (1990) *Immunology,* **71,** 38–45.
16. Lau, R.K.W., Hill, A., Jenkins, P., Caun, K., Forster, S.M., Weber, J.N., McManus, T.J., Harris, J.R.W., Jeffries, D.J. and Pinching, A.J. (1992) *Int. J. STD AIDS,* **3,** 261–266.
17. Clavel, F., Guetard, D., Brun-Vezinet, F., *et al.* (1986) *Science,* **233,** 345–346.
18. Pepin, J., Morgan, G., Dunn, D., Gevao, S., Mendy, M., Gaye, I., Scollen, N., Tedder, R. and Whittle, H. (1991) *AIDS* **5,** 1165–1172.
19. Kong, L.I., Lee, S. and Kappes, J.C. (1988) *Science,* **240,** 1525–1529.
20. Biggar, M. (1990) *International Registry of Seroconverters.* AIDS incubation in 1891 HIV seroconverters from different exposure groups.
21. Blanche, S., Mayaux, M-J., Rouzioux, C. *et al.* (1994) *New Engl. J. Med.,* **330,** 308–312.

Chapter 4

Clinical aspects of HIV

4.1 Introduction

HIV mainly infects cells bearing the CD4 surface molecule, which acts as a specific receptor for the viral envelope protein, gp120. Such cells are found predominantly within the immune system and include T-helper lymphocytes, monocytes and antigen-presenting cells. However, there are also CD4+ cells within the central nervous system [1], these being the microglial cells which are of monocyte or macrophage lineage. These cells can be productively infected by HIV *in vitro*, and *in vivo* there is evidence of an HIV-induced cytopathic effect since syncytia-like, multi-nucleated cells are seen in the brains of HIV-infected individuals. This CD4-defined tissue tropism explains the major pathological effects of HIV, which are immunodeficiency and neurological disease. However, HIV may also cause damage at sites where CD4 is not expressed. This may be a result of direct infection of CD4− cells, in which low levels of viral replication can occasionally occur [2] (*Table 4.1*), or due to infiltration of tissue by HIV-infected lymphocytes and macrophages, which release toxic viral proteins and/or pro-inflammatory cytokines.

4.2 Diagnosis of HIV infection

In adults, the antibody response is the most reliable way to make a diagnosis of HIV infection. HIV culture requires specialist facilities and is relatively insensitive. PCR is 'oversensitive', being able to detect very small copy numbers of genetic material, including minute contaminating levels of RNA or DNA, and therefore may give false positive results. Viral p24 antigen can be routinely measured in the serum, but is only transiently detectable immediately prior to seroconversion, remaining at undetectable levels in most individuals until the development of HIV-related illnesses.

Anti-HIV antibodies are usually detected by enzyme-linked immunosorbent assay (ELISA) or Western blot techniques. Although HIV-1 and

Table 4.1: Cells infectable by HIV

CD4+ cells	
Immune system	T-helper lymphocytes
	Monocytes/macrophages
	Dendritic cells
Brain	Microglia
Skin	Langerhan's cells
Gastrointestinal tract	Human colorectal cells[a]
Liver	Kuppfer cells[a]
CD4− cells	
Immune system	CD8 lymphocytes
	B-lymphocytes[a]
Brain	Glial cells
	Astrocytes
	Human neuroblastoma cells[a]
	Retina cells
Lung	Fibroblasts
Kidney	Epithelial cells
Gastrointestinal tract	Columnar and epithelial cells
	Enterochromaffin cells
Liver	Human hepatoma cells[a]
Bone marrow	Stem cells

Adapted from ref. [2].
[a]Infectable *in vitro*.

HIV-2 are only around 50% homologous at the genetic level, many of the envelope epitopes are conserved between the two. Kits are now widely available which detect both types of antibodies. The main limitations of the antibody test are as follows:

(1) It will not become positive until 6 weeks to 3 months after exposure; therefore, there is a 'window' where diagnosis is not possible using this test;

(2) It is of no use in the diagnosis of HIV infection in the first months of life in children born to infected mothers, since the maternal IgG antibody passes the placenta, meaning that all infants are 'HIV positive' (the diagnosis of HIV in infants is discussed in Section 4.12.3);

(3) It will not detect infection in hypogammaglobulinemic individuals;

(4) Tests may not differentiate between individuals who are truly infected and those that have been vaccinated with HIV in vaccine trials (although not a major problem currently, this could be so if vaccines prove to be effective and have a widespread use in groups at risk).

These limitations apply to only a fraction of individuals being tested and, as long as they are recognized, the antibody test remains a robust means of diagnosing HIV infection that can be performed in routine laboratories on large numbers of serum, heel prick or salivary samples.

HIV testing has important implications for the individual concerned. HIV is a potentially fatal disease, so if someone is tested positively then their sexual partners and children may also be infected; if they are negative they may have to wait some weeks to eliminate the 'window' period. For these reasons, pre- and post-test counseling is essential to the psychological health of the client.

4.3 Classification of HIV infection

The clinical staging system of the Center for Disease Control (Atlanta, USA) is widely used in developed countries for the classification of HIV infection (*Table 4.2*). It largely relies on accurate and definitive diagnosis of infection, which make it less applicable to poorer nations. The three stages of early HIV infection, seroconversion illnesses, asymptomatic carriage and PGL, fall into the 'A' category. During this stage the levels of CD4+ lymphocytes are often normal, although their function, in particular the proliferative response to soluble antigens, may already by affected.

Table 4.2: Classification of HIV infection

	Clinical Categories		
CD4 count	(A) Acute infection, asymptomatic or PGL	(B) HIV-related conditions[a], not A or C	(C) AIDS-defining diagnoses[a]
≥500 μl^{-1}	A1	B1	C1
200–499 μl^{-1}	A2	B2	C2
< 200 μl^{-1}	A3	B3	C3

Based on the Center for Disease Control (CDC) criteria for adults/adolescents ≥13 years age, with proven HIV infection.
[a]See *Tables 4.3* and *4.4*.

As disease progresses and the CD4 count falls, individuals develop minor illnesses which may reflect direct HIV damage (e.g. thrombocytopenia or peripheral neuropathy) or early immunological dysfunction (e.g. oral candidiasis, hairy oral leukoplakia, seborrhoeic dermatitis or constitutional symptoms). These are the 'B' category conditions listed in *Table 4.3*, and sometimes termed AIDS-related complex or 'ARC'. These conditions are not life threatening, but are a sign that the HIV disease is progressing and therefore may predict the development of full-blown AIDS with major opportunist infections and tumors (*Table 4.4*). Patients may pass from any of the clinical stages to AIDS, but to date no individuals have shown a regression of HIV-related disease back to a clinically and immunologically stable state (*Figure 4.1*). The major clinical presentations of AIDS are discussed below.

Table 4.3: Examples of category B conditions

Bacillary angiomatosis
Candidiasis, oropharyngeal, (thrush)
Candidiasis, vulvovaginal, frequent or poorly responsive to therapy
Cervical dysplasia (moderate to severe)/cervical carcinoma *in situ*
Constitutional symptoms, fever (38.5°C) or diarrhea for > 1 month
Hairy oral leukoplakia
Herpes zoster (shingles) involving at least two distinct episodes or more than one dermotome
Idiopathic thrombocytopenic purpura
Listeriosis
Pelvic inflammatory disease
Peripheral neuropathy

Table 4.4: AIDS-defining diagnoses (category C)

Candidiasis of bronchi, trachea or lungs
Candidiasis of esophagus
Cervical carcinoma, invasive
Coccidioidomycosis, disseminated or extrapulmonary
Cryptococcosis, extrapulmonary
Cryptosporidiosis, chronic intestinal (> 1 month's duration)
CMV disease (other than liver, lymph node or spleen)
CMV retinitis
Encephalopathy, HIV-related
Herpes simplex: chronic ulcers (> 1 month's duration); or bronchitis, pneumonitis or esophagitis
Histoplasmosis, disseminated or extrapulmonary
Isosporiasis, chronic intestinal (> 1 month's duration)
KS
Lymphoma, Burkitt's
Lymphoma, immunoblastic
Lymphoma, primary of brain
MAC or *Mycobacterium kansasii*, disseminated or extrapulmonary
Mycobacterium tuberculosis, any site
PCP
Bacterial pneumonia, recurrent
Progressive multifocal leukoencephalopathy
Salmonella septicemia, recurrent
Toxoplasmosis of brain
Wasting syndrome, due to HIV

4.4 Features of disease in the immunocompromised host

Immunodeficiency is characterized by impaired handling of infectious agents, leading to 'opportunist infections', defined as (1) diseases caused by usually non-pathogenic organisms or (2) unusually severe diseases caused by known pathogens, and 'opportunist tumors', which are usually

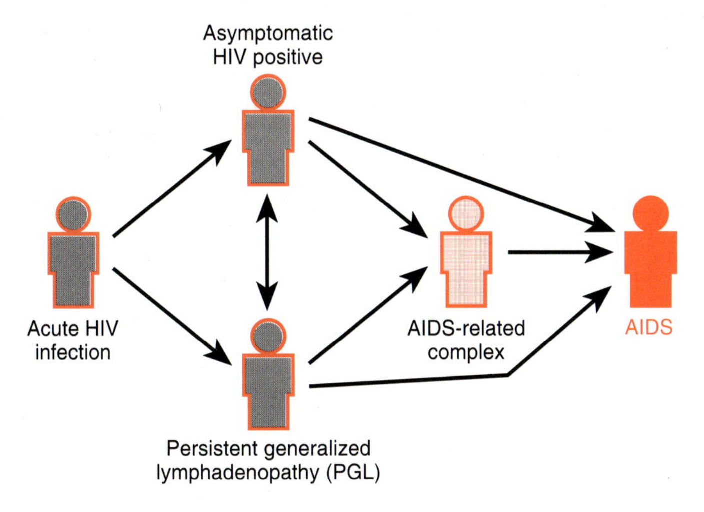

Figure 4.1: Interaction of the clinical stages of HIV infection.

due to the uncontrolled activity of oncogenic viruses. In general, defects in different arms of the immune response lead to different patterns of opportunist disease (*Table 4.5*). Phagocytes (neutrophils and monocytes), together with the humoral factors (complement and antibody), protect against extracellular organisms (predominantly bacteria). The cell-mediated immune response controls intracellular pathogens and consists of:

(1) CTLs which recognize foreign antigen complexed with the class I histocompatibility antigen and eradicate virally infected cells;

(2) Macrophages which ingest and kill many fungi, protozoa and parasites as well as the facultative intracellular bacteria, including salmonella and mycobacteria;

(3) NK cells which are involved in tumor surveillance.

The cellular response is orchestrated by the CD4 lymphocyte, which is destroyed in HIV infection; therefore, it is the intracellular infections that predominate. A similar pattern of infection is seen in other conditions involving the disruption of the cellular immune system, such as the congenital immunodeficiencies, Di George syndrome and SCID, Hodgkin's disease and iatrogenic immunosuppression due to systemic corticosteroids, chemotherapy, cyclosporin A and FK507 used to control rejection of transplanted organs.

4.5 Effect of immunosuppression on clinical presentation

Diagnosis of infection in the immunocompromised host may be difficult. First, the clinical symptoms and signs are lacking or altered, due to the defective inflammatory response; for example, patients with PCP may be

Table 4.5: Characteristic infections/tumors associated with deficiencies of different parts of the immune system

	Defects		
	Neutrophil	Antibody/ complement	Cell-mediated immunity
Viruses		Echovirus	Polio
			Herpes
			Herpes simplex virus 1 & 2
			Varicella zoster virus
			EBV
			CMV
			Adenovirus
			Polyoma: BK, JC
			Papilloma
		Mycoplasma	
Bacteria	*Staphylococcus epidermidis*	*Staphylococcus aureus*	Salmonella
	Staphylococcus epidermidis	Pneumococcus	
	Escherichia coli	*Hemophillus influenzae*	Listeria
	Klebsiella	Moraxella	
	Pseudomonas	Meningococci	
	Anaerobes		
			Mycobacteria
			Tuberculosis
			MAC
			Nocardia
Fungi	Candida (disseminated)		Candida (muco-cutaneous)
	Aspergillus		Cryptococcus
			Histoplasma
Protozoa			*Pneumocystis carinii*[a]
			Toxoplasma
			Cryptosporidia
Tumors			NHL (B-cell lymphoma)
			Kaposi's sarcoma
Other			GVHD

[a]RNA analysis suggests this organism to be more related genetically to fungi, but phenotypically it behaves as a protozoan.

breathless and hypoxic, but the chest examination and radiograph may show minimal or no abnormality. This suggests that normally the shadows seen in chest X-ray in pneumonitis and the signs of consolidation are as much due to the influx of inflammatory cells as to the organisms themselves. Secondly, the immune system cannot be used to diagnose infection. For example, CD4 lymphocytes are required to induce granuloma formation in mycobacterial infections. Their failure to respond in HIV infection means that the usual histological tell-tale sign of tuberculosis is often absent or poorly formed, and Mantoux tests cannot be used to diagnose the disease reliably. In addition, B-cells are dysfunctional and they respond poorly to new antigen challenge. Polyclonal activation of memory B-lymphocytes leads to high levels of antibody to previously encountered antigens, even if the infections are no longer active. Thus, the serological diagnosis of infections is highly unreliable (see *Table 4.6*).

The atypical presentation, failure of indirect tests to diagnose infection, and the common occurrence of multiple pathogens, means that direct examination of the fluid or tissue for organisms is required. In such immuno-compromised patients it is essential that a rapid diagnosis is made, as the earlier that disease is treated, the greater the likelihood of response.

Table 4.6: Clinical features and management implications of immunodeficiency in HIV infection

Effect of immunodeficiency	Clinical implication
Opportunist infection	Infection with usually non-pathogenic organisms Unusually severe/disseminated infection with known pathogens
Opportunist tumors	Due to uncontrolled oncogenic viruses
Recurrent infections	Long-term suppressive treatment may be required
Multiple concurrent infections	If a patient does not respond to appropriate treatment then another disease process must be suspected
Lack of usual symptoms and signs due to failure of inflammatory response	Look directly for organisms or antigens in tissue or body fluid
Inability to use immune response to diagnose infections	Direct identification of organisms required
Failure of immunological control of allergic disease	Reactivation of atopic disease, high rate of drug reaction

4.6 Pattern of presentation of opportunist disease in HIV infection

A knowledge of which diseases are likely to present at different stages of HIV infection not only improves the likelihood of making the correct diagnosis when patients present with problems, but also allows appropriate preventative (prophylactic) treatment to be targeted.

4.7 Effect of degree of immunodeficiency

Different infections tend to emerge at different disease stages (*Figure 4.2*); early in HIV infection, when immunodeficiency is slight, the infections

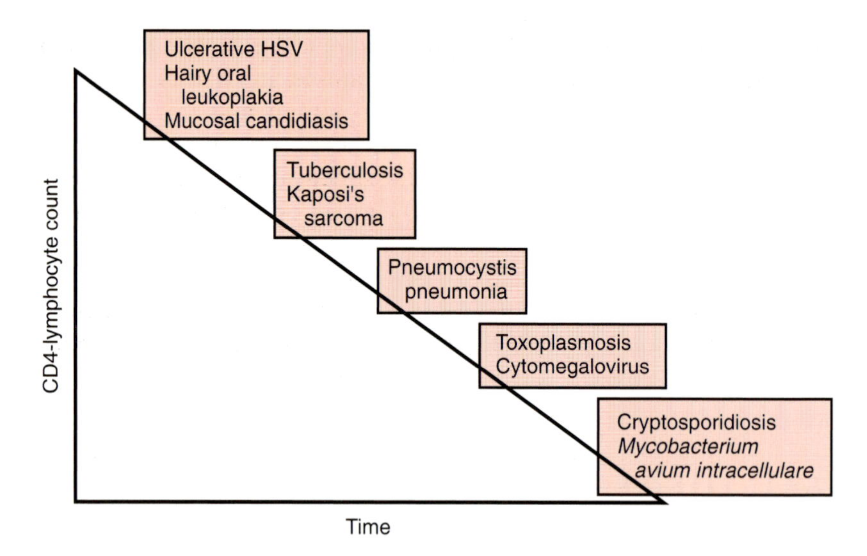

Figure 4.2: The effect of the level of immunosuppression on the development of opportunist disease in HIV infection. High-grade pathogens such as herpes simplex virus (HSV), candida and tuberculosis are able to cause disease in immunocompetent hosts; therefore, only a minor impairment of immunity (as judged by CD4 count) is needed for these infections to emerge. Conversely, low-grade pathogens, such as disseminated *Mycobacterium avium intracellulare*, only cause illness in patients with severe immunodeficiency.

that occur are with relatively high-grade pathogens that cause disease even in the immunocompetent host. Examples are herpes simplex virus, varicella zoster (shingles), wart virus infection, mucocutaneous candidiasis or tuberculosis, the main difference being that these infections are more severe or persistent than would normally be expected. Kaposi's sarcoma may also develop in patients whose immune systems are still relatively well maintained. In advanced disease, disseminated infections with organisms of low pathogenicity can occur, especially *Mycobacterium avium* complex (MAC), CMV, cryptosporidia and microsporidia, most of these being confined to patients with a CD4 count less than 100 cells μl^{-1}. Lymphoma shows two patterns, with central nervous system involvement usually occurring late in HIV-related disease, and having a poor prognosis, whereas lymphoma outside the CNS, particularly nodal lymphoma, may occur in relatively early HIV-related infection, and respond well to therapy.

4.8 Effect of microbiological environment

The majority of infections are reactivations of a previously acquired disease that has remained latent. Therefore, the diseases that emerge when immune deficiency develops represent the infections to which the person has been exposed in the past. The 'microbiological environment' varies from country to country (*Table 4.7*). For example, in African countries, cryptococcus and tuberculosis are extremely common, and are likely to be the first conditions to reactivate with increasing immunodeficiency in AIDS. PCP, which is prevalent in the developed world and is found in 50% AIDS cases at presentation, is only rarely seen in African AIDS cases, unless they have lived outside Africa. Likewise, histoplasmosis is a common infection in mid-west USA, often acquired asymptomatically; it is only when individuals become immunosuppressed that the disease is manifested. Disseminated *Penicillium marnefii* infection appears largely confined to individuals who have lived or traveled in the Far East.

Table 4.7: Effect of microbiological environment on pattern of disease presentation in AIDS

Frequency	Europe	USA	Africa
Very high	PCP	PCP	Mycobacterium *tuberculosis*
Very high	CMV disease	CMV disease	Esophageal candidiasis
High	Cerebral toxoplasmosis	Atypical mycobacteria	Cryptococcal meningitis
High	Atypical mycobacteria	Cerebral toxoplasmosis Histoplasmosis and other disseminated mycoses	
	KS in homosexual men	KS in homosexual men	KS in heterosexual men and women
Low	NHL	NHL	
Low	Cryptococcal meningitis	Cryptococcal meningitis	PCP, CMV

4.9 Direct effects of HIV infection

4.9.1 Nervous system

HIV infection is associated with a wide variety of neurological disorders which appear to be directly mediated by HIV (those which are a result of opportunist infection are discussed in Section 4.10). Encephalopathy is a common manifestation. This is associated with diffuse atrophy of grey and white matter on computerized tomographic (CT) or magnetic resonance imaging (MRI) scans, with ataxia and dementia clinically. Postmortem studies using *in situ* hybridization have shown localized areas of HIV infection and replication within the brain [3]. However, the extent of neurological damage is out of proportion to the relatively small viral load, and it is likely that secreted viral proteins, such as gp120 (which has been shown to be directly neurotoxic in animal studies) are also to blame. Involvement of toxic HIV proteins has also been suggested from clinical studies, as treatment with the anti-HIV drug zidovudine not only reduces the incidence of encephalopathy, but also consistently causes improvement in encephalopathic features, even in those who have irreversible cerebral atrophy. HIV also causes disease at other neurological sites, as listed in *Table 4.8*. The majority of patients with neurological disease also have immunodeficiency, but in a few individuals these features are isolated, and it has been suggested that some strains of HIV are predominantly 'neurotropic'.

Table 4.8: Direct neurological effects of HIV

Neurological disease	Clinical features
Encephalopathy	Cerebral atrophy
	Trunkal ataxia
	Cognitive impairment
Peripheral neuropathy	Sensory loss
	Axonal abnormalities
	Lower > upper limbs
Autonomic neuropathy	Autonomic nerve damage in small bowel biopsies
	Diarrhea
	Postural hypotension
Myelopathy	Motor and sensory loss
	Vacuolar myelopathy on histology
Myositis	Myalgia
	Proximal muscles > distal
	Raised creatine phosphokinase levels
Retinitis	Microvascular retinopathy
	Cotton wool spots
	Microaneurysms and hemorrhages

4.9.2 The gastrointestinal tract

The rectum as a site for acquisition of HIV infection. Penetrative anal intercourse between men is a major route of transmission of HIV in the developed world. The mucosa is thin and easily torn and virus within the lymphocyte-containing semen could directly infect colonic cells which are susceptible to HIV. However, the infectability of these cells is relatively low, and it is more likely that rectal lymphoid aggregates, containing CD4 lymphocytes and follicular dendritic cells, are the targets.

HIV enteropathy. HIV enteropathy describes the chronic pathogen-negative diarrhea and malabsorption sometimes associated with marked weight loss that occurs in HIV-positive individuals. The pathological findings are of subtotal villous atrophy and extensive damage to autonomic fibers in the small bowel [4]. It has recently been shown that 30% or more of the lymphocytes in the lamina propria are HIV infected in AIDS. These cells actively secrete cytokines, in particular IL-1β and

TNFα. These 'pro-inflammatory' molecules not only cause inflammation, but may also increase gut permeability and may cause the structural abnormalities described above. This condition appears to be more common in African individuals, and in these countries 'slim' disease is synonymous with AIDS [5]. This may reflect a true difference in HIV pathogenesis, although the possibility that the diarrhea is due to infectious agents cannot be excluded.

4.10 Diseases resulting from the immunodeficiency

4.10.1 Pneumocystis carinii *pneumonia (PCP)*

Until the use of preventative therapy, PCP was the most common opportunist infection in patients with AIDS, occurring in 50% as their presenting diagnosis, and in 60% at some stage in their disease. *Pneumocystis carinii* is classified morphologically as a protozoan, although it has some genetic features that are closer to the fungi. It is ubiquitous in many countries, and up to 50% of healthy populations in such areas will have been exposed to the organism by adult life. Normally it is non-pathogenic, but in the immunocompromised host it causes a potentially life-threatening pneumonitis. Much of our knowledge about pneumocystis comes not from HIV-infected individuals, but from outbreaks that occurred in malnourished children in eastern Europe and the Middle East in the 1950s, and from patients who are immunosuppressed by leukemia, chemotherapy or intensive immunosuppression for organ or bone marrow transplant.

Presentation. The characteristic history of PCP is of a progressive non-productive cough (or productive of a small amount of white frothy sputum), increasing shortness of breath on exertion, increased respiratory rate, exercise desaturation, hypoxia and fever. Pulmonary function tests show a decrease in transfer factor for carbon monoxide. Patients with a CD4 count of less than 200 μl^{-1} are particularly at risk, with 30% developing PCP within 18 months. Smokers also appear to be at greater risk. The timecourse may be prolonged over several weeks (not all infections in the immunocompromised occur rapidly). Examination of the lungs may be normal or reveal fine crepitations. The classical radiological findings are bilateral peri-hilar interstitial shadowing, with sparing of the apices and bases. However, a normal chest radiograph does not exclude the diagnosis.

Diagnosis and management. As the cough is minimally productive and the burden of infection lies within the alveoli, bronchoscopy with broncho–alveolar lavage (BAL) or transbronchial biopsy is usually required for detection of the cysts. An alternative method of obtaining the material involves inhalation of hypertonic saline, which leads to exudation of alveolar fluid that can be expectorated with the aid of physiotherapy, an 'induced sputum' sample. This can yield positive results, but the sensitivity is low (50% compared with 95% for BAL).

Treatment is with high doses of co-trimoxazole or intravenous pentamidine. The use of high-dose systemic corticosteroids in the initial treatment phase of those with severe PCP ($PO_2 < 70$ mmHg) has led to a marked improvement in survival. It is proposed to act by reducing the adult respiratory distress syndrome that may occur as a reaction to the death of organisms on initiation of therapy.

Prophylaxis. The risk of PCP has been greatly reduced by the use of prophylactic therapy with lower dose co-trimoxazole three times weekly or nebulized pentamidine for those at risk (those with a CD4 count < 200 μl^{-1} undergo 'primary prophylaxis', while those who have already had PCP are given 'secondary prophylaxis'). The main long-term complication of PCP is pneumothorax secondary to post-infection cysts. Extrapulmonary PCP occurs rarely, particularly in those on nebulized pentamidine prophylaxis [6].

4.10.2 CMV disease

CMV is a member of the herpes virus family. Disease in HIV infection is due to reactivation of a previously acquired infection; therefore, it is particularly common in homosexual men, as 95% are carriers of CMV compared with 30–40% of the heterosexual population. In AIDS the main presentations are with retinitis and colitis. Pneumonitis, nervous system involvement and esophageal disease occur less frequently [7]. CMV generally occurs in the severely immunocompromised (CD4 counts < 100 cells μl^{-1}).

CMV retinitis. Patients may complain of visual loss, 'floaters' or retro-orbital pain. However, the disease may be asymptomatic if the periphery of the retina is involved. The appearance is characteristic with exudates and hemorrhages occurring along the line of the blood vessels (*Figure 4.3*). This has been described as 'crumbly cheese and tomato ketchup' or 'pizza pie' change. It is also described as a 'wildfire' retinitis as it spreads rapidly, destroying all in its path. The retina cannot heal itself, and as the disease is frequently bilateral if left unchecked, patients will therefore be blinded unless managed aggressively.

The drugs ganciclovir or foscarnet are successful in controlling the infection. These agents suppress viral replication, but cannot eradicate it; therefore, long-term maintenance is required to keep the disease under control. The major late complication is retinal detachment occurring at sites of scarring.

Gastrointestinal tract disease. CMV colitis presents with abdominal pain, rebound tenderness (often in the left iliac fossa), and non-bloody diarrhea.

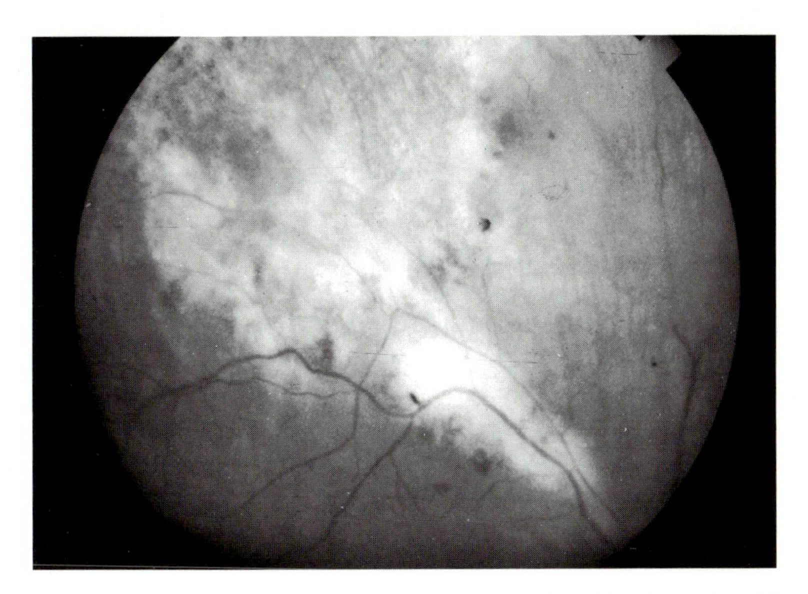

Figure 4.3: CMV retinitis. The picture shows retinal blood vessels with white fluffy exudate running alongside. The darker blotches mixed in with the exudate are hemorrhages indicative of active disease. The retina cannot heal itself even when the infection is controlled, and an area of scarring will remain.

Severe cases may progress to toxic megacolon and perforation. Diagnosis is made by detection of the typical cytomegalic inclusion bodies on histology of biopsy samples. CMV can also cause disease at other sites in the gut; esophageal infection presents with solitary ulcers, usually at the lower end of the esophagus; small bowel infection usually consists of multiple ulcers. CMV may also invade the biliary tract, and together with cryptosporidium and microsporidium is a cause of AIDS-related sclerosing cholangitis.

Neurological disease. CMV encephalitis appears to be becoming more common as more severely immunocompromised patients are surviving longer. The infection may be predominantly peri-ventricular, presenting with confusion and short-term memory loss with an abnormal electro-encephalogram (EEG). The other presentations are a polyradiculopathy or transverse myelitis. Examination of cerebrospinal fluid typically shows a neutrophil leukocytosis. Culture of the fluid may reveal CMV, and PCR has been shown to be both sensitive and specific for CMV-related neurological disease in this situation. Response to therapy can be seen, but it often slow possibly due to poor penetration of anti-CMV agents into the cerebrospinal fluid.

4.10.3 Cerebral toxoplasmosis

As is the case for many infections with HIV, toxoplasmosis is due to a reactivation of previously quiescent disease, and only extremely rarely a result of a new infection. Therefore, although antibody levels cannot be used to diagnose the infection, they can be used to identify those who may be at risk since those who have had toxoplasma in the past will have positive serology throughout their lives. The risk of developing toxoplasmosis if sero-positive is 30% during the period of AIDS. Therefore, the possibility of prophylaxis is now being tested in clinical trials.

The main features are multiple cerebral abscesses, which show typical ring enhancement after contrast has been given on CT or MRI scanning (*Figure 4.4*). Patients present with focal neurological signs; for example, weakness of one side of the body, difficulty in speaking or fits, with confusion, headache and fever. Response to treatment with sulfadiazine and pyrimethamine is generally good, but does not eradicate the infection. Therefore, lifelong maintenance is required to keep the infection controlled once induction treatment is completed. Surprisingly, retinal toxoplasmosis is uncommon in this group.

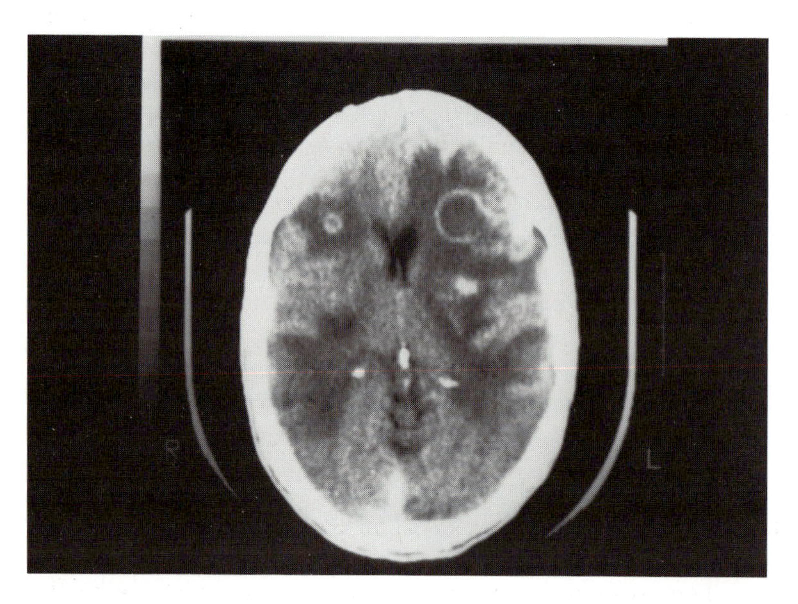

Figure 4.4: A CT scan of the brain showing multiple ring enhancing abscesses in a patient with cerebral toxoplasmosis. With appropriate treatment all these lesions will completely disappear.

4.10.4 Mycobacterial disease

Mycobacterial infection falls into two categories: (1) *Mycobacterium tuberculosis*; and (2) the atypical mycobacteria, MAC, and a few other environmental species. These present at different stages of HIV infection and cause different patterns of disease (*Table 4.9*).

Table 4.9: Comparison of *Mycobacterium tuberculosis* and *Mycobacterium avium intracellulare* complex (MAC) infection in AIDS patients

Tuberculosis	MAC
Lung main site of disease	Gastrointestinal tract main site
CD4 may be normal	CD4 count very low: < 100 mm^{-3}
Good response to treatment	Poor response to treatment
May show granuloma formation	No granuloma formation

Tuberculosis. *M. tuberculosis* is a high-grade pathogen that causes disease even in individuals with an apparently normal immune system. Therefore, it is a major pathogen to those with impaired cellular immunity, and in areas where both HIV and tuberculosis are common in the community (e.g. Africa, Asia and deprived inner cities) this infection is a major cause for concern. In the USA the previously sustained decline in tuberculosis rates was reversed in 1984 as the HIV epidemic took hold, and in other countries it was a resurgence of tuberculosis that brought HIV within the community to light. With the increase in cases has come the increase in multidrug resistance, emphasizing the importance to public health of ensuring compliance to full treatment regimes. Presentation in the immunocompromised is similar to that in immunocompetent patients; that is, the lung is the most common site, although the infection is more likely to be disseminated, with miliary disease in the lung, liver, pleural, pericardial and meninges.

Tuberculosis in AIDS usually responds well to standard combination chemotherapy using drugs such as isoniazid, rifampicin, pyrizinamide and ethambutol, but since it is clear that therapy never completely eradicates the organisms treatment programs are followed by lifelong chemoprophylaxis. In HIV-positive individuals from areas with high rates of tuberculosis, the chances of reactivation are very high, but can be significantly reduced by the use of isoniazid as primary prophylaxis.

MAC. The atypical mycobacteria are minimally pathogenic and only cause severe disease when there is marked immunosuppression. Unlike tuberculosis, where many of the infections appear to be a reactivation of latent disease, MAC infection is likely to be acquired. The main route of infection is through the gastrointestinal tract, and most of the localized disease occurs in the small bowel causing malabsorption and diarrhea, in the large bowel causing diarrhea, or in the liver causing hepatomegaly (enlargement of the liver) and abnormalities of liver function (raised levels of alkaline phosphatase). The lymph nodes draining these sites are also involved and may become very enlarged causing abdominal pain. In addition, there is usually disseminated disease causing high fevers, and bone marrow suppression with anemia. The organism is slow growing and these symptoms usually occur over the course of some weeks. The organisms are typically resistant to standard drugs, but two agents, rifabutin (a rifamycin similar to rifampicin) and a clarithromycin (a macrolide with a good macrophage concentration), have good activity, usually in combination with ethambutol. However, treatment is generally only able to suppress the organism load, and is therefore aimed at palliating symptoms.

Cryptococcal meningitis. Cryptococcus is an encapsulated yeast that is carried in bird droppings, particularly from pigeons. It is acquired through inhalation of the organisms, and may cause a pneumonitis. More commonly, the presentation is with meningitis and it is likely that this is due to dissemination of a new or previously acquired infection. Even when severe infection is present, patients with cryptococcal meningitis frequently present without the typical signs of neck stiffness and photophobia, and despite the presence of many organisms in the cerebro-spinal fluid, there may be no associated white cells. Diagnosis is made on identification of the organisms directly, or by measurement of cryptococcal antigen levels. Patients show a good response to the anti-fungal agents amphotericin B and fluconazole. However, the organism is usually not eradicated and maintenance therapy is required. Relapses are more difficult to treat, and resistant organisms are becoming more common. The most important long-term sequelae of this infection, as of other meningitides, is the development of hydrocephalus due to the removal of impaired cerebrospinal fluid.

Cryptosporidiosis and microsporidiosis. Cryptosporidium is a protozoan that was previously well known to vets, since it infects fowl and cattle. There were only a few reported cases in humans before 1980, but this changed when it was found to be a frequent cause of diarrhea in AIDS patients. Further investigation has now shown that in fact it is a cause of self-limiting diarrhea in travellers and occasionally follows contamination of water supplies.

The organism is extracytoplasmic, but fuses to the enterocyte wall of cells in the small bowel, and induces an intense secretory diarrhea of up to 10 l per day. There has been little success in treatment to date, most of the therapies being supportive to maintain hydration and nutrition. However, a non-absorbable aminoglycoside, paromomycin has given encouraging results symptomatically, and in a few cases has markedly reduced the cyst load.

Microsporidium is another small bowel pathogen that has also been associated with diarrhea in HIV infection. Its pathogenic role has yet to be established, as some individuals may carry the organism asymptomatically. Albendazole is useful in the treatment of some species of microsporidium.

Both these organisms are associated with AIDS-related sclerosing cholangitis, a condition where the bile ducts become scarred and narrowed. Both these organisms only tend to cause severe disease in the very immunosuppressed host (with CD4 counts $< 100\ \mu l^{-1}$).

4.10.5 Kaposi's sarcoma

Kaposi's sarcoma (KS) is caused by a proliferation of endothelial cells of vascular or lymphatic origin. It is a polyclonal rather than a monoclonal population, and is therefore not a true malignancy. Histologically, slit-like spaces lined by spindle-shaped tumor cells are formed. Red blood cells become trapped in these channels and give the tumor its characteristic purple–red color. KS has been recognized since the middle of the nineteenth century, being found in individuals from central Africa and elderly men often of Jewish origin. In these populations KS is an indolent, slowly progressive disease, usually confined to the skin: so-called 'endemic KS'. In AIDS, 'epidemic KS', the tumor is much more aggressive. However, apart from being cosmetically unsightly, often the spots do not cause major problems when confined to the skin. Local radiotherapy, freezing or ablation by laser can be very effective. Internal lesions in the gastrointestinal tract or lung require more aggressive treatment with chemotherapy, as they can cause obstruction. Zidovudine and IFN-α have some effect on KS lesions.

There has been intense interest in the pathogenesis of KS. The close association with immunodeficiency suggested that an infectious agent may be involved [8]. This has been fueled by the observation that KS is now being observed in HIV-negative homosexual men who have a normal immune system [9], suggesting that either the tumor cells or the etiological agent is being transmitted sexually in this group. KS was previously rare in heterosexual AIDS cases outside Africa, but is now being observed. However, the majority of the cases have had sexual contact with someone who also has KS.

4.10.6 Lymphomas and other malignancies

The majority of lymphomas occurring in AIDS are B-cell-derived non-Hodgkin's lymphoma (NHL), similar to those observed in other immunocompromised hosts. There is a strong association with EBV infection, with nearly 100% of brain NHLs expressing latent membrane protein or other markers of infection [10]. Indeed, the use of PCR to detect EBV in the cerebrospinal fluid has proved a useful diagnostic tool for nervous system lymphoma in HIV infection.

The tumors are commonly 'extra nodal'; that is, not within lymph nodes, particularly in the very immunocompromised patients. The most common sites are gut, lung and brain.

These lymphomas are usually sensitive to chemotherapy and radiotherapy, and these treatments are useful for extracerebral disease.

Intracerebral lesions are difficult to treat; chemotherapy does not penetrate the brain easily and the response to radiotherapy is poor.

In addition to frank lymphomas, HIV-positive patients also develop syndromes of 'EBV-driven lymphoproliferation', as seen in iatrogenically immunosuppressed people. The proliferation is polyclonal (and so not malignant) and may respond to the anti-EBV drug acyclovir. However, it is likely that these lymphoma-like lesions progress to NHL with time.

HIV-infected children develop what appears to be another proliferation driven by EBV: lymphoid interstitial pneumonitis (LIP). This does not appear to have malignant potential.

Several other tumors appear to be slightly more common in those with HIV infection. Of these, the strongest evidence is for squamous cell carcinoma of the anus; like NHL, there may be an underlying viral cause in the oncogenesis, the human papilloma virus being the most likely candidate in this particular carcinoma.

4.10.7 Allergic disease

There is a marked increase in allergic drug reactions in HIV-positive patients, in particular to sulfonamides, but also to ciprofloxacin, grizeofulvin, isoniazid, pyrizinamide and rifampicin. Skin reactions and abnormal liver function are the most common features. This appears to be a feature of many immunodeficiencies, and although this may appear a paradox at first, it is not altogether surprising. It is now established that CD4+ lymphocyte subsets with characteristic cytokine secretion patterns are central in the control of allergic responses. The Th_1 subgroup of CD4 cells produces IFN-γ and reduces IgE production, while the Th_2 subgroup produces IL-4 and enhances IgE production. In HIV infection, the ability to secrete IFN-γ in response to mitogen or antigen is poor; this loss would be expected to allow the escape of IgE-secreting B-cells and the development of allergic disease. This has been found to be the case; those with a history of atopic disease frequently develop recrudescence of symptoms or a worsening of persistent disease with progressive immunodeficiency. Anecdotally, recombinant IFN-γ treatment has been demonstrated to control these symptoms [11].

It is of interest that some conditions appear to be very rare in HIV-infected populations, suggesting that an intact cellular immune system is required for their pathogenesis. Crohn's disease and sarcoidosis are not reported, whereas ulcerative colitis may still occur. The similar transmission of hepatitis B and HIV means that dual infections are common in homosexual men and intravenous drug users. The immunodysregulation of HIV infection makes patients less able to clear hepatitis B, and HIV-positive individuals do not respond to IFN-α immunomodulation [12]. Therefore, there are many HIV-positive persistent carriers. However, HIV-positive individuals are less likely to have active disease and scarring on histology [13].

4.11 Change in disease pattern in AIDS

Since the first AIDS cases were treated in the early 1980s, there have been considerable advances in the prevention and treatment of opportunist infections, as well as the development of specific therapies for the underlying HIV infection. This has been associated with a marked decline in the incidence and mortality, due to PCP in particular. The survival of patients with AIDS has doubled over this time, to a median of about 22–24 months in most centers in developed countries [14]. However, as the deaths due to opportunist infections have declined, the mortality due to tumors has increased (*Table 4.10*). This reflects the observation that although KS and lymphoma may respond well to treatment initially, with time they become resistant. It is therefore the prevention and therapy of these tumors, as well as the treatment of the difficult opportunist infections, that requires improvement to enhance quality and quantity of life.

Table 4.10: Changing pattern of causes of mortality in AIDS (values are given in percentages)

Cause of death	1982–83	1984	1985	1986	1987	1988	1989
PCP	33	29	35	46	40	27	3
Other OI	67	57	47	43	27	34	49
KS	0	14	12	11	23	27	32
NHL	0	0	6	0	10	12	16

Taken from ref. [14].
OI, opportunistic infections.

4.12 Pediatric HIV infection

Throughout the world, the majority of HIV infection is transmitted by heterosexual intercourse. This means that as the epidemic continues, more and more women, and the children that they subsequently bear, will be infected. The adage that children are not just small adults is highly applicable to HIV infection. Diagnosis cannot be made in the same way, and the presentation of HIV-related diseases is different in the developing neurological and immunological systems (*Table 4.11*).

4.12.1 HIV infection in pregnancy

It is debatable whether or not HIV infection progresses more rapidly during pregnancy. It is theoretically a risk, as the fetus is an allogeneic

Table 4.11: Differences between adult and pediatric presentations of HIV infection

Adult	Child
Mainly cell-mediated immune deficiency	Cell-mediated and antibody deficiency
Outcome from PCP good (< 10% mortality)	Outcome from PCP poor (> 90% mortality in first year of life)
Lymphoid interstitial pneumonitis rare	Lymphoid interstitial pneumonitis common
KS common	KS rare
CD4 counts good predictor of prognosis	CD4 counts poor predictor of prognosis
HIV in nervous system leads to dementia	HIV in nervous system leads to loss of milestones

stimulus, which is known to activate lymphocytes and HIV. In addition, there is the physiological immunosuppression which allows 'half-foreign' fetal tissue to survive. However, as discussed on p. 38, in practice there may be no adverse effects on the mother. From the child's perspective, transmission is most likely to occur if the mother is at an advanced stage of HIV disease. Therefore, if a woman wishes to have children, it is better that she does so at the earliest stage possible. A placebo-controlled study of zidovudine given to HIV-positive mothers antenatally and during delivery, and to the child immediately after birth, has shown that the use of this anti-retroviral can significantly decrease the transmission rate. The development of a positive therapeutic approach in this situation may change the balance for many women towards having an HIV test in pregnancy.

The management of HIV during pregnancy is confused by the decline in CD4 counts that occurs in all women at this time, regardless of HIV. It is unknown whether this reflects a real decrease in immune function, or merely a redistribution of the CD4+ lymphocytes. However, standard prophylactic regimes for PCP are used if the CD4 count drops below 200 μl^{-1}.

4.12.2 Breast feeding

The rate of transmission of HIV is significantly increased by breast feeding. The virus is found within the fluid and lymphocytes of the milk, and HIV-infected wet nurses have passed infection on to babies they are feeding. In developed countries mothers are therefore advised against breast feeding. However, in third world countries the situation is different, as the mortality from bottle feeding is around 15 times higher than that from breast feeding, due to difficulties in sanitary preparation of the feeds.

In these countries breast feeding is recommended to all mothers, regardless of HIV status.

4.12.3 Diagnosis of HIV infection in infants

HIV-infected infants can become ill very rapidly. The earliest cases of PCP (*Figure 4.5*) occur before 3 weeks of age, and as over 90% cases of PCP in the first year of life are fatal, it is important that infants are monitored for infections and given prophylaxis with co-trimoxazole. If the infant fails to thrive and HIV-related neurological deterioration occurs, the child may respond to anti-retrovirals (*Figure 4.6*). However, all of these interventions carry some risk and it is important to be able to find out as soon as possible which children are infected. This is not as straightforward as in the adult. All infants born to HIV-infected mothers will be HIV-antibody positive due to the passive transfer of maternal antibodies across the placenta. In the first few months of life there is no difference in antibody titer between infants who are and are not infected, and the IgM response used in the diagnosis of many other acute viral infections in infancy is unhelpful. Thus, serology is only helpful in the second year of life, when

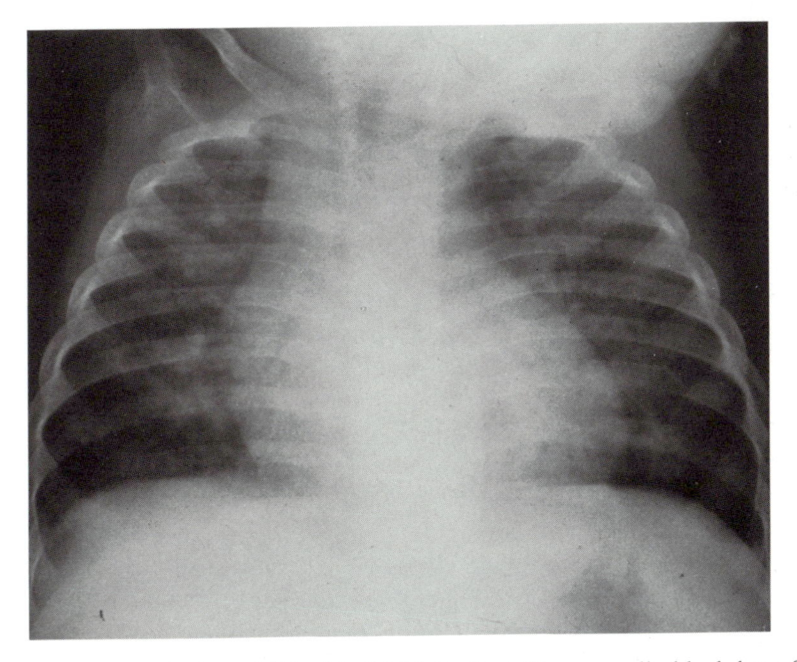

Figure 4.5. PCP in a child with HIV infection. The normally black lung fields show multifocal patchy white shadowing typical of an interstitial pneumonitis. The mortality from PCP in the first year of life is over 90%, possibly because it represents a primary infection in this group, whereas in adults disease is due to re-activation or re-infection in pre-exposed individuals. Photograph courtesy of Dr Sam Walters, St Mary's Hospital, London.

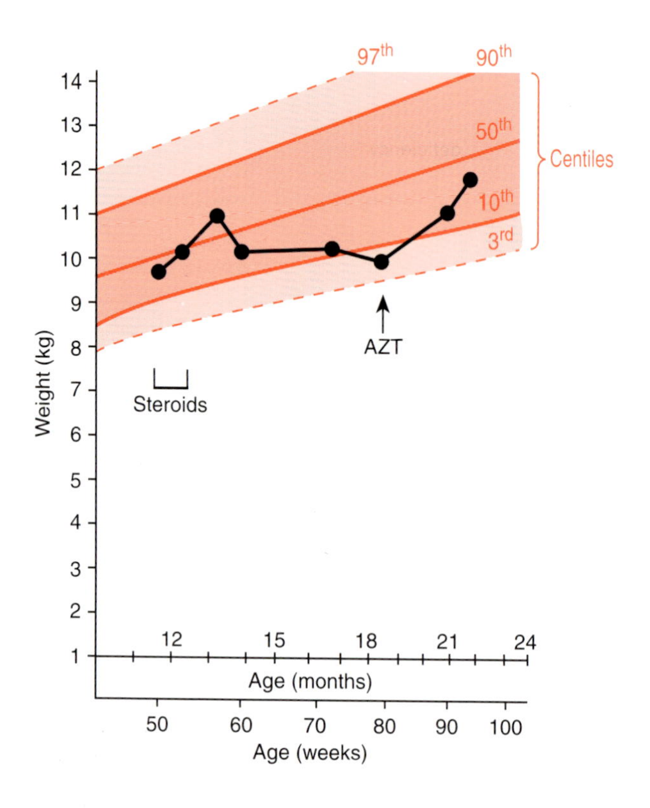

Figure 4.6: The effect of zidovudine on growth slowing in a child with HIV infection. Courtesy of Dr Di Gibb, Great Ormond Street Hospital, London.

uninfected children lose their maternal Ig and become antibody negative for HIV. Likewise, the markers for immunodeficiency, in particular the close association of CD4 numbers and susceptibility to opportunist infections, does not hold true in the pediatric setting. Total lymphocyte and CD4 cell numbers are very high at birth, showing a progressive decline to reach adult levels around adolescence (*Figure 4.7*). Infants can develop PCP with a CD4 count in the 500–1000 range per mm^3. The percentage of CD4 cells is much more stable at around 40–45% throughout childhood and adult life, and this may prove a better prognostic marker than absolute numbers. p24 antigen is very rarely raised in the infected infant in the first few months of life.

Other markers such as hemoglobin, β_2 microglobulin and Ig levels may be of help in adding further evidence for, or against, infection. However, these are non-specific markers and serial analyzes are needed to obtain trends, all of which takes time. The best tool available for making an early diagnosis is HIV culture. In certain specialized laboratories, HIV can be cultured from 95% of infected individuals, regardless of disease stage. HIV can be isolated from infected infants in the first few weeks of

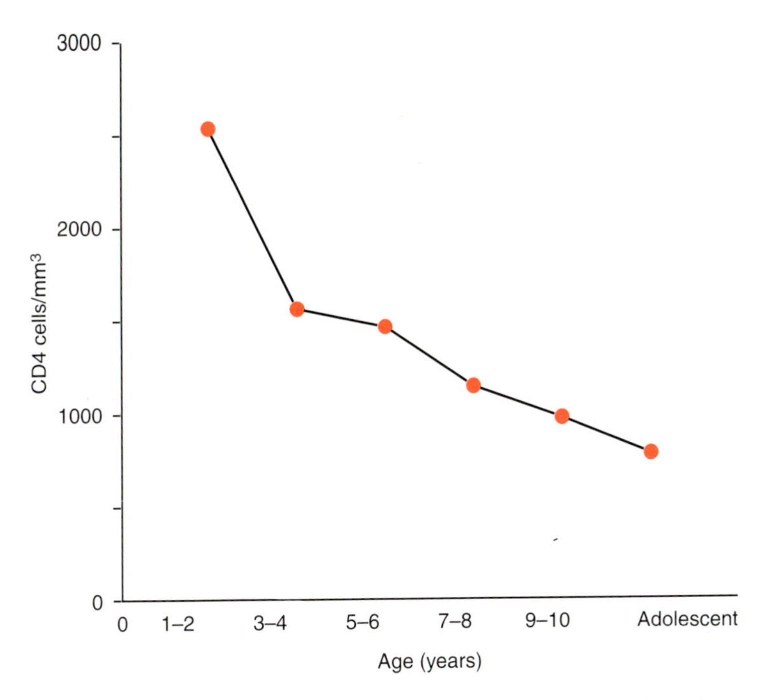

Figure 4.7: CD4 counts are highest at birth, dropping consistently to reach adult levels by adolescence. This physiological change makes it difficult to interpret decreases in CD4 counts in children. In addition, children can develop opportunist infections despite CD4 counts that would be normal in an adult. Adapted from ref. [15].

life. Repeatable negative cultures make the diagnosis of HIV infection unlikely. PCR can also be used, but the difficulty in interpreting the status of the small number of infants who are repeatedly culture negative but PCR positive makes it a more difficult result to analyze.

Until the status of the child is known, they can be given co-trimoxazole as PCP prophylaxis. Standard vaccinations for tetanus, diphtheria, measles, mumps and rubella can be given. Bacillus Calmette Guerin (BCG) is contraindicated until HIV status is known, as disseminated infection with the vaccinating organism can occur. Live polio vaccine (Sabin) is probably safe, but use of the inactivated (Salk) version is currently recommended.

The difficulty in the diagnosis of HIV infection in infants leads to medical dilemmas in terms of the use of vaccinations, prophylaxis and anti-retroviral agents, and emotional distress for the parents who may have to wait many months to be certain of the status of their child.

References

1. Maddon, P.J., Dalgleish, A.G. and McDougal, J.S. (1986) *Cell,* **47,** 333–348.
2. Cheng-Mayer, C. (1990) *AIDS,* **4** (Suppl. 1), 49–56.

3. Koenig, S., Gendelman, H.E. and Orenstein, J.M. (1986) *Science,* **233,** 1089–1093.
4. Griffin, G.E., Miller, A., Batman, P., Forster, S.M., Pinching A.J., Harris, J.R.W. and Mathan, M.M. (1988) *AIDS,* **2,** 379–382.
5. Serwadda, D., Mugerwa, R.D. and Sewankambo, N.K. (1985). *Lancet,* **2,** 849–852.
6. Northfelt, D.W. (1989) *Lancet,* **ii,** 1454.
7. Tyms, A.S., Taylor, D.L. and Parkin, J.M. (1989) *J. Antimicrob. Chemother.* **23,** 89–105.
8. Beral, V., Peterman, T.A., Berkelman, R.L. and Jaffe, H. (1990) *Lancet,* **335,** 123–128.
9. Friedman-Kien, A.E., Saltzman, B.R., Cao, Y., Nestor, M.S., Mirabile, M., Li, J.J. and Peterman, T.A. (1990) *Lancet,* **335,** 168-169.
10. MacMahon, E.M.E., Glass, J.D., Hayward, S.D., Mann, R.B., Becker, P.S., Charache, P., McArthur, J.C. and Ambinder, R.F. (1991) *Lancet,* **338,** 969–973.
11. Parkin, J.M., Eales, L-J., Galazka, A.R. and Pinching, A.J. (1987) *Br. Med. J.,* **294,** 1185.
12. Brook, M.G., McDonald, J.A., Karayiannis, P., Caruso, L., Foster, G., Harris, J.R.W. and Thomas, H.C. (1989) *Gut,* **30,** 1116–1122.
13. Goldin, R.D., Fish, D.E., Hay, A., Waters, J.A., McGarvey, M.J., Main, J. and Thomas, H.C. (1990) *J. Clin. Pathol.,* **43,** 203–205.
14. Peters, B.S., Beck, E.J., Coleman, D.G., Wadsworth, M.J.H., McGuinness, O., Harris, J.R.W. and A.J.P. (1991) *Br. Med. J.,* **302,** 203–207.
15. Bofill, M., Janossy, G., Lee, C.A., MacDonald-Burns, D., Philips, A.N., Sabin, C., Timms, A., Johnson, M.A. and Kernoff, P.B.A. (1992) *Clin. Exp. Immunol.,* **88,** 243–252.

Chapter 5

Treatment and monitoring of HIV infection

5.1 Introduction

The treatment of HIV-related immunological and neurological disease involves the control of HIV replication by specific anti-viral drugs, boosting the host's own immune response to HIV, dampening down any potentially harmful results of the immune response, and reconstituting the defective cell-mediated immunity. Thus, a broad range of approaches are being tried. This in some way reflects the uncertainty regarding the pathogenesis of this infection. The management of the specific opportunist infections and tumors that arise because of the immunodeficiency are covered in Chapter 4. Here, we will focus on the treatment of the underlying HIV infection.

5.2 Specific anti-retroviral therapy

The ultimate goal for the HIV-infected person would be the eradication of HIV infection. However, this is unlikely to be achieved as the virus integrates into the host chromosome, and therefore all infected cells would have to be removed. This may be possible in renewable tissue, such as lymphoid cells, but not within the nervous system, where cells cannot be replaced. However, if the virus could be suppressed to the level where it is not damaging the host or continuing to infect new cells, then the disease process may be halted. The virus can potentially be inhibited in three ways: (1) by preventing binding or fusion to the target cell; (2) by inhibiting replication and release; and (3) by inducing non-infective particles (*Figure 5.1*).

5.3 RT inhibitors

The discovery that a retrovirus causes AIDS led to a worldwide search for an agent that would control the infection. There was much information

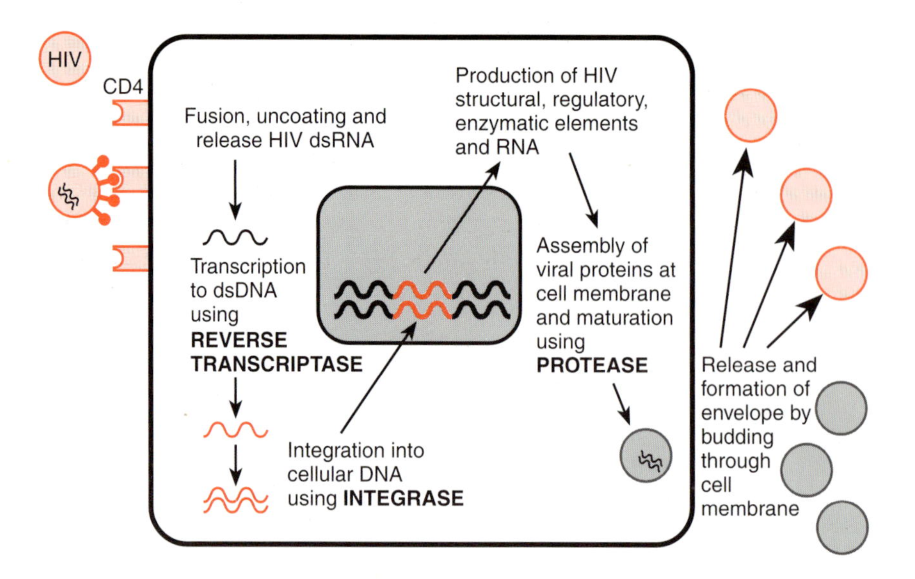

Figure 5.1: Targets for anti-HIV agents. All stages of HIV replication, from initial binding to the cell, through reverse transcription and integration to eventual budding, release and maturation, are potential targets for therapy. Antibodies to the HIV envelope, block binding (neutralizing antibodies), gp120 can also be blocked with a soluble form of its CD4 ligand, or conversely the CD4 receptor itself can be blocked with anti-CD4 antibodies. Polyanions also block the binding of HIV to the cell surface, possibly by steric hindrance. IFNs (α and β) inhibit viral replication and may reduce infectability of cells generally. Enzyme inhibitors of HIV can block the RT, integrase and protease leading to failure to establish infection, failure of integration and failure of maturation, respectively. 'Intracellular immunization' is being approached using antagonists to the regulatory molecules of HIV (including Tat), or by blocking the binding sites for these products on the genome. ds, double-stranded.

already available on the RT of the virus from previous work on HTLV-1. This enzyme is an attractive target for therapy: (1) it acts at two stages in the viral pathway (being required for initial provirus establishment within the cell and for viral replication); and (2) the enzyme has no closely related cellular homolog (the nearest being the DNA α, β and γ polymerases), and treatment would therefore be relatively specific.

5.3.1 Nucleoside analogs, AZT, ddI and ddC

Several drugs that had been developed to combat cancer had anti-RT activity, and were therefore 'taken off the shelf' for testing against the new virus. In the mid-1980s, the nucleoside analog, azidothymidine (zidovudine; AZT) was found to inhibit viral replication in laboratory cultures, and was the first drug to be used in large placebo-controlled studies.

5.3.2 Mechanism of action

The agent is given as the pro-drug, which is metabolized within the cell to the active form, AZT 5′ triphosphate. This is a competitive inhibitor of thymidine 5′ triphosphate, acting as a chain terminator by preventing the formation of 5′–3′ phosphodiester bonds. Thus, the virus cannot make a DNA copy and integration cannot occur. It is relatively specific for the viral RT, but as it is a nucleoside analog, it also has affinity for the mammalian polymerases, especially the cellular γ polymerase found in mitochondria. This is likely to be the cause of the mitochondrial myopathy found with prolonged AZT usage.

5.3.3 Clinical uses

AZT was found to improve survival and reduce infection rate in patients with AIDS and advanced ARC [1] (*Table 5.1*). Although it was clear that AZT did not cure HIV-related disease, the survival of patients with AIDS appeared to be prolonged by about 10–12 months (from a median of approximately 12 months to 22 months). This was a significant step forward, and the use of AZT in advanced disease has been confirmed from other studies. Treatment tends to lead to an increase in CD4 count and a decrease in p24 antigen levels in such patients (*Figure 5.2*), and may cause improvement in lymphocyte function as well as absolute numbers [2]. The other areas in which AZT has been shown to be of use is in the treatment and prevention of diseases directly mediated by HIV (thrombocytopenia, HIV encephalopathy and neurological disorders), and more recently in the reduction of maternal–fetal transmission. Interesting results from a number of studies have shown that the concomitant use of acyclovir (an anti-herpes drug) with AZT improves clinical outcome. The mechanism for this is unknown. There is much less information available to guide its use in the pediatric setting. However, it does appear to have a role, and in some infants can halt the progression of neurological disease and reverse growth retardation (*Figure 4.6*).

Table 5.1a: Effect of zidovudine on clinical outcome in patients with ARC or AIDS

	Placebo $N=137$	Zidovudine $N=145$
Deaths	19	1
Episodes of opportunist infection	45	24

Taken from ref. [1].

Table 5.1b: Hematological toxicity of zidovudine in patients with ARC or AIDS

	Placebo	Zidovudine
Anemia (<7.5 g/l^{-3})	4%	21%
Neutropenia (<500 mm^{-3})	2%	16%

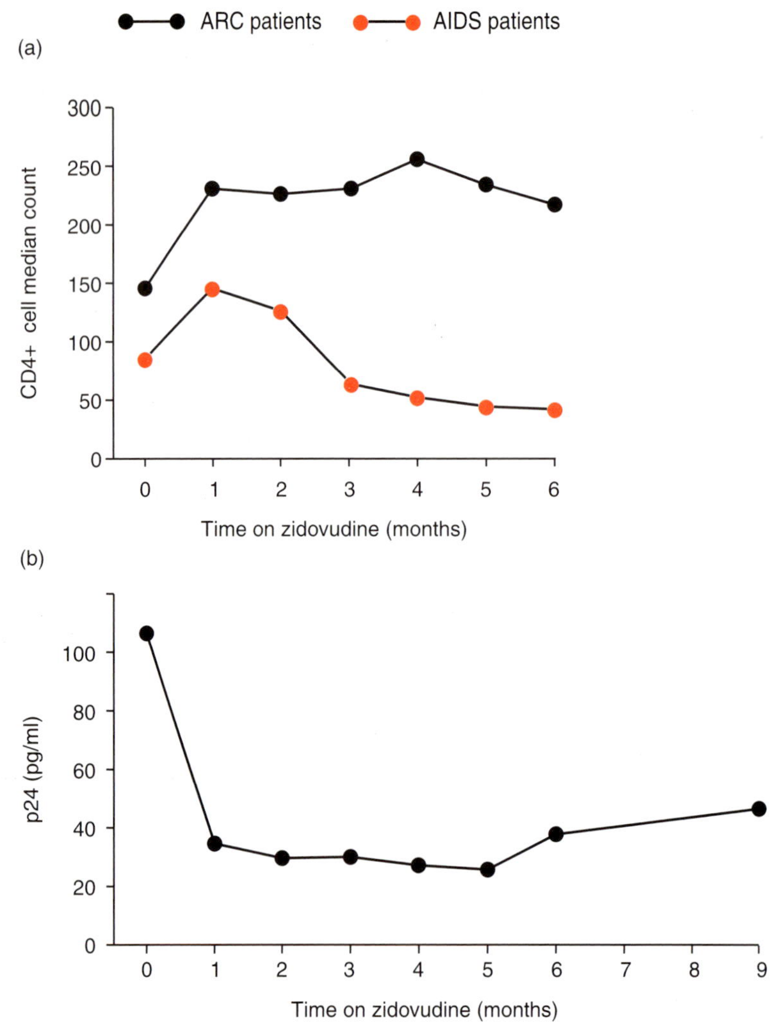

Figure 5.2: Changes in CD4 count (a) and p24 antigen level (b) in patients on zidovudine treatment. Patients characteristically show an elevation in CD4 count after 4 weeks of treatment. In those with AIDS, this is only maintained for a short period, whereas in those with less advanced immunodeficiency (ARC) it may persist. The serum p24 antigen declines rapidly with therapy and remains depressed for several months in most patients. Eventually it will start to increase again, possibly due to the development of zidovudine-resistant viruses.

Following the success of this drug in the mid-1980s, several other RT (polymerase) inhibitors with similar mechanisms, such as zalcitabine (dideoxycytidine; ddC) and didanosine (dideoxyinosine; ddI), were developed and are now licensed in many countries. Other analogs, D4T and 3TC, are also being investigated. These appear to show useful effects, but have been less vigorously tested clinically than AZT; their use in, for example, neurological disease and platelet disorders has not been confirmed.

The limitations of these agents became clear through the initial studies and follow up of patients on long-term treatment. First, they display degrees of toxicity. Initially this is mainly due to the effects on the rapidly dividing cells, in particular the bone marrow with AZT usage. ddI and ddC have different toxicity profiles, as shown in *Table 5.2*. Secondly, HIV can become resistant to these drugs. HIV mutates both between and within individuals, including elements of the RT enzyme. Resistant strains of HIV are commonly found after 6 months or more of AZT treatment. This is due to point mutations in the viral genome, especially at positions 67, 70, 215 and 219. These mutations lead to a loss of affinity for the anti-retroviral drug, but despite the changes, the transcriptase enzyme is able to maintain its physiological function.

As the toxicity profiles are different, alternating or combination regimes of the nucleoside analogs can be successful in avoiding major side-effects. In addition, as initially there is only rare crossresistance between these agents, and as strains may revert to sensitivity if the drug is not used for a period, alternating regimes of AZT and ddI and ddC may allow longer periods of efficacy.

Table 5.2: Main side-effects of currently used anti-retroviral agents

AZT	ddI	ddC
Early (1–2 weeks):	Acute pancreatitis (1%)	Peripheral neuropathy
headache and nausea	Hyperamylasemia	
Medium (4–6 weeks):	Diarrhea	
bone marrow suppression,	Hypokalemia	
especially anemia	Dry mouth	
Late (> 12 months):	Peripheral neuropathy	
myopathy (mitochondrial)		
Gastrointestinal disturbances		

5.3.4 Use of zidovudine in early HIV infection

Following the success of zidovudine in patients with advanced disease, further trials took place in individuals with earlier stages of HIV infection, with the rationale that it might be possible to prevent the development of HIV-related disease. The history of the investigation of zidovudine illustrates the importance of conducting appropriate placebo-controlled trials. Because the pressure to find a cure for the disease was so great, many of the anti-retroviral studies which followed were either not placebo controlled, used 'surrogate' endpoints (changes in the CD4 count or p24 antigenemia), or used rather 'soft' clinical endpoints. Such trials suggested that zidovudine may prevent disease progression when used at an asymptomatic phase of HIV infection. These trials generally used small numbers of patients and/or had only a short follow up time. However, a long-term, placebo-controlled European study, which investigated the effect of early compared with deferred zidovudine treatment in 1727 asymptomatic HIV-positive individuals, has put these results into question. In agreement with other trials, this study showed that use of zidovudine was associated with a higher CD4 count. After 1 year of treatment with AZT, some effect was shown on reducing the development of minor symptoms. However, this benefit was completely lost after 3 years and there was no reduction in the development of AIDS (18% in both groups). There was also no effect on mortality, being 8% in the zidovudine-treated group and 6% in the placebo-controlled group. It is likely that AZT does control HIV progression but only for a limited time, and that the emergence of viral resistance undermines any early benefit. This does not mean that anti-retroviral drugs are of no use in early infection, but that improved regimes will have to be developed.

5.4 Non-nucleoside RT inhibitors

Other inhibitors of RT have been developed acting through different mechanisms. The main drugs in this group are the bis-hetero-arylpiperazine (BHAP) and tetrahydroimidazo-benzodiazepin-thione (TIBO) compounds, including nevirapine. These agents are potent inhibitors of HIV replication acting on a different site on the RT to nucleoside analogs and therefore potentially of use in synergistic combinations. They appear to be allosteric inhibitors of the enzyme. Some are like benzodiazepines, but lack the sedative or addictive effects, and are well tolerated, apart from maculo–papular skin rashes. All the compounds bind to the same site on the RT, and are therefore all susceptible to the several RT substitutions that can confer viral resistance. Unfortunately, resistance develops within 6 weeks of starting treatment, so as single agents they have no practical use. Combination studies are underway with nucleoside analogs.

5.5 Protease, integrase and Tat inhibitors

HIV has other specific enzymes that can be targeted. Inhibitors to the integrase that allows integration of the viral genome into the host genetic material, and to the protease that is necessary for formation of mature infectious virus, are all in various stages of testing. Inhibitors of Tat and manipulation of the other viral regulatory elements is also possible.

5.6 Other approaches

Soluble CD4 can now be produced and *in vitro* inhibits the binding of HIV to cells by blocking gp120 on the viral envelope. *In vivo,* it has a very short half-life, even with genetic manipulation, and no consistent effects have been noted. Antibodies to the virus or host cells, including anti-idiotypic antibodies, have been investigated with the aim of preventing virus binding to its target. The results have been variable. Other agents can block binding non-specifically. These are a variety of polyanion substances including heparin and dextran sulfate. A difficulty with all these approaches is that the molecules need to coat the virus or host cell in order to be protective. It appears that much HIV infection of cells occurs by cell–cell transmission, not involving free virion. The approaches described would be unlikely to affect this mechanism.

5.7 General approaches to maintenance of health

Given the limitation on specific anti-HIV drugs, it is important that factors that can adversely affect the immune system are controlled as much as possible, so that any damage induced by HIV is not compounded. The particular areas where this can be positively approached are as follows.

5.7.1 Nutrition

Protein-calorie malnutrition is a cause of significant cellular immunodeficiency, which may contribute to the immunocompromise in AIDS. There are many reasons why a patient with HIV infection may eat less and also burn up more energy than normal (*Table 5.3*). The individual causes of these need to be corrected if possible. High protein and calorie food supplement drinks can increase calorie intake significantly. These can be given through a nasogastric tube in the short term or via a semi-permanent tube going into the stomach directly from the abdominal wall, which is inserted via a gastroscope (percutaneous endoscopic gastrotomy). Occasionally patients with severe malabsorption require intravenous (parenteral) nutrition.

Table 5.3: Factors affecting nutrition in HIV infection

Event affecting nutrition
Decreased consumption
Pain: mouth or esophageal infection
Anorexia or nausea and vomiting:
infections
medications
Small stomach:
Multiple KS lesions
Weakness, lethargy, depression
Decreased absorption
HIV enteropathy
Small bowel infection:
Mycobacterium avium intracellulare
Cryptosporidium
Increased metabolism
Fever

5.7.2 Stress

There is evidence that psychological factors can affect immune responses. For example, studies of students before examinations shows marked changes in T-cell numbers. This is not surprising in the light of the close interaction of the neurological, endocrinological and immunological systems. There are also data from studies on patients with cancer that particular psychological attitudes to the disease at diagnosis affected ultimate prognosis; those with either a positive approach or a denial response fare better than those with a negative response. Therefore, supporting the individual psychologically if required is likely to improve the clinical outcome, although this is very difficult to measure objectively.

5.7.3 Reducing lymphocyte activation

Lymphocyte activation is a potent stimulus to HIV replication within the immune system. Therefore, it is expected that episodes of intercurrent infection which activate lymphocytes would enhance HIV progression. For these reasons, infections should be diagnosed and treated rapidly, and preventative therapies used for commonly recurring diseases such as herpes simplex and oral candidiasis.

5.7.4 Differences in HIV strains

There is some evidence that different strains of HIV have varying potential to cause disease. There is a case where an HIV-infected blood donor has remained well for several years. The six people who became infected by receiving his blood have also been long-term survivors, apart from one patient who was given systemic corticosteroids. Babies infected *in utero* from mothers who have high titers of virus (presumably the fast-replicating strains seen in late disease) become ill significantly more rapidly than those who contract infection from mothers with low titers of virus (presumably a slow-growing strain). Therefore, it may be beneficial to limit as much as possible the numbers and types of strains of HIV which are acquired. Safe sexual practices may be recommended even when both partners are HIV positive.

5.8 Monitoring HIV infection

Monitoring HIV infection involves regular clinical examination (to determine any clinical signs of progression, such as oral candida or hairy leukoplakia, as well as to detect early signs of infections so that these can be treated effectively), and laboratory tests of markers of HIV activity and progression. Careful monitoring allows appropriate decisions to be made about the use of anti-retroviral drugs and prophylaxis of opportunist infections. Although not all individuals wish to have these interventions, most will wish to be aware of their health in order that they can make necessary decisions about their lives.

5.8.1 Immunological monitoring

CD4 lymphocyte counts. The CD4 count is the most useful surrogate marker of HIV-related disease. This gives some index of prognosis; that is, patients with counts less than 200 per mm^3 have high progression rates to AIDS over short periods of time, and benefit from PCP prophylaxis and possibly AZT treatment. However, there are some individuals who develop major opportunist disease despite maintained CD4 counts, and conversely those who run low CD4 counts and yet remain healthy. This lack of specificity and sensitivity is probably due to the test measuring numbers of cells, not their function. Caution needs to be exercised in the interpretation of the results, since many factors besides HIV lead to alteration in CD4 levels (*Table 5.4*). Serial values are useful, but it is rarely of value to perform these more that once every 3 months, as the background 'noise' obscures any meaningful interpretation.

Functional lymphocyte tests (proliferation to anti-CD3 antibody and poke weed mitogen) have been developed and are being analyzed for use in disease monitoring and as a marker of efficacy of anti-retroviral agents. Currently, these tests remain a research tool, as they are technically demanding and not yet applicable to routine laboratories. Markers of CD8 activation are also being assessed.

Table 5.4: Factors affecting CD4 count

Physiological	
Age	Highest at birth, progressive decrease with age
Diurnal variation	Peak at 23.00, trough at 11.00
Pregnancy	Decreases
Exercise	Increases
Smoking	Increases
Pathological	
Congenital immunodeficiency	Decreases all T-cell subsets
DiGeorge syndrome, SCID	Decrease in CD4 cells
Idiopathic CD4 depletion	Decrease in CD4 cells
Tuberculosis	Decrease in CD4 cells
Sarcoidosis	Decrease in CD4 cells
Autoimmune disorders, Sjogren's	Lymphopenia with predominant decrease in CD4 cells
Systemic corticosteroids and other immunosuppressives, post BMT	Lymphopenia with predominant decrease in CD4 cells

β_2 Microglobulin and neopterin. β_2 Microglobulin is a marker of lymphocyte activation while neopterin marks macrophage activation. Both rise with progressive HIV infection, but β_2 microglobulin tends to discriminate better for progression to early symptomatic disease (ARC) whereas neopterin indicates the development of progression to the more severe disease (*Figure 5.3*). The marked overlap between groups of patients at different clinical stages for both these markers means that alone they are of little use in clinical monitoring. They may be useful markers for efficacy in therapeutic trials.

5.8.2 Virological monitoring

p24 Antigen and antibody. The p24 antigen is often used as a marker for viral activity, and its presence is associated with progression of disease (*Table 5.5*). Anti-HIV agents that show evidence of clinical efficacy lead to a decline in p24 antigen levels. Therefore, it is likely that p24 antigen has some clinical relevance. However, up to 50% of patients with AIDS may be antigen negative, and this proportion is even higher in African groups. The use of acid dissociation to enable p24 antigen hidden in immune complexes to be detected increases the reliability of this test. p24 Antigen inversely correlates with p24 antibody and it is very unusual to have both detectable in the serum. It appears that the B-cells that secrete the p24 antibody are lost during disease development, and this may precede the emergence of p24 antigen in the serum.

Viremia is better quantified by culture (tissue culture infectious dose) or by PCR analysis. However, these techniques are only applicable to peripheral blood samples as these are the only cells that are easily accessible for monitoring. Whether the results reflect what is happening in the whole body is uncertain.

Phenotype. The two HIV phenotypes, SI and NSI, have been shown to correlate with disease progression in studies of small numbers of patients. This test requires virus isolation facilities and is not available in routine laboratories.

Table 5.5: p24 Antigen as prognostic marker in HIV infection

Clinical group	% positive for p24 antigen
HIV antibody negative (risk matched)	0
HIV + asymptomatic	4
PGL	25
ARC	56
ARC progressing to AIDS	75
ARC remaining stable	33
AIDS	70

Taken from ref. [3].

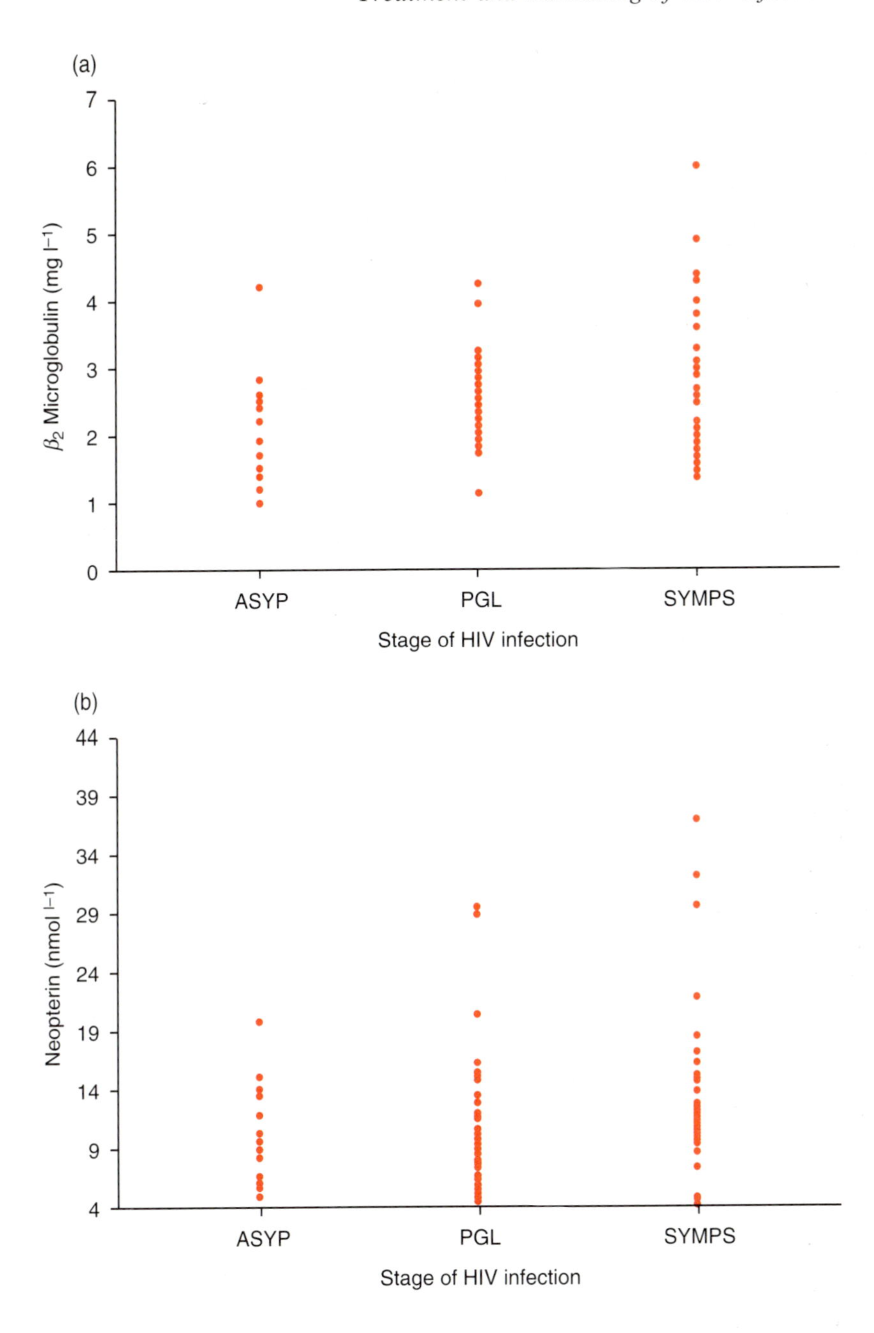

Figure 5.3: β_2 Microglobulin and neopterin levels at different stages of HIV infection [asymptomatic (ASYMP), PGL and symptomatic disease (SYMPS), including ARC and AIDS]. Both show a progressive rise as disease develops. Data courtesy of Dr Mark Gompels, Medical College of St Bartholomew's Hospital, London.

References

1. Fischl, M.A., Richman, D.D., Grieco, M.H., Gottlieb, M.S., Volberding, P.A., Laskin, O.L., Leedom, J.M., Groopman, J.E., Mildvan, D. and Schooley, R.T. (1987) *New Engl. J. Med.,* **317,** 185–191.
2. Nye, K.E., Knox, K.A. and Pinching, A.J. (1991) *AIDS,* **5,** 413–417.
3. Kenny, C., Parkin, J.M., Underhill, G., Shah, N., Burnell, R., Osborne, E. and Jeffries, D. (1987) *Lancet,* **294,** 1185.

Chapter 6

Immunomodulation and approaches to HIV vaccines

6.1 Immunomodulation

Given the current limitations on anti-HIV drugs, it seems logical to attack the problem from other directions as well. The second part of this chapter will focus on the development of vaccines to prevent HIV infection. For those who already have the infection, improved adjuvants and the production of many immunological cytokines by recombinant technology means that directed immunomodulation is now possible. This could be used in two ways in established HIV infection: to improve the host's own immune response to the virus; and to restore the defective cellular immunity.

6.2 Therapeutic immunization

Therapeutic immunization consists of vaccination of a person already infected with an organism to enhance the immune response. In the case of HIV, this approach has not been widely investigated, largely because of the uncertainty of the immunopathogenesis of HIV, and the fear that an 'autoimmune' process may be aggravated. Phase I trials have begun using a p24 virus-like protein construct which has proved immunogenic and shows promise for future immunotherapy.

6.3 Passive immunotherapy

Passive immunotherapy consists of donating elements of the specific immune response (usually antibody) to the infected individual. There have been a number of studies, mostly uncontrolled, that have suggested that specific antibody to envelope or core proteins may have some beneficial effect on CD4 counts and clinical features of HIV-related disease. However, how much is due to the specific antibody and how much to the general effects of giving Ig to patients who are known to have a functional hypogammaglobulinemia, remains uncertain.

6.4 Interferons and interleukins

The technology to produce large amounts of cytokines by recombinant genetics was developed very close to the time of the discovery of AIDS. Therefore, it is not surprising that there have been many studies of these agents in HIV infection, and much of the knowledge about these substances has come from these groups. In terms of immunoreconstitution, the two most logical candidates are IL-2 and IFN-γ. Both these lymphokines are normally produced by CD4-helper lymphocytes and help to amplify immune responses. In HIV infection the production and regulation of these lymphokines is defective.

IFN-γ has several roles. It activates macrophages to kill intracellular pathogens such as mycobacteria, fungi and protozoa, it upregulates MHC class II expression on antigen-presenting cells to stimulate presentation to CD4 cells and generate the necessary lymphocyte proliferation, and it activates NK cells to virucidal and tumoricidal activity.

All these functions are defective in HIV-infected individuals, but can be improved by adding IFN-γ *in vitro* and *in vivo*. Initial clinical trials showed some encouraging effects with immunomodulation *in vivo* as well as a reduction in HIV antigenemia [1]. It is of interest that as with many cytokines there is a therapeutic window for IFN-γ, and that pushing the dose too high can lead to opposite effects, with a resulting increase in immunosuppression. Placebo-controlled trials are underway to determine whether IFN-γ will give protection from opportunist infections.

The role of IL-2 is to activate lymphocytes, in particular cytotoxic T-cell 'killer' cells. It has shown some activity in cancer treatment, when lymphocytes are removed from the patient (the population containing within it tumor-specific cytotoxic cells) and expanded with IL-2 *in vitro*. These lymphokine-activated killer 'LAK' cells are then injected back into the patient, and target the tumor. The lack of IL-2 production was thought to underlie the susceptibility to viral infections and tumors in AIDS patients, and therefore trials were constructed to test the efficacy of this agent in reconstitution of the immune response. Initial reports did not indicate efficacy and there was concern that IL-2 would activate lymphocytes and thereby enhance HIV replication.

6.5 HIV vaccines

The road leading to the development of a successful vaccine against HIV contains many obstacles. One major problem has been the lack of an animal model; HIV does not replicate to high titer in chimpanzees, neither do they contract disease. The simian immunodeficiency virus (SIV) of primates, while causing infection, does not mimic HIV infection although some preliminary data suggest that the macaque may be susceptible to HIV-1. The great variability of the envelope protein of HIV poses yet another problem: there is no guarantee that a vaccine raised against one HIV-1 isolate would be effective against another isolate of HIV-1. The cost of a successful vaccine would also be important. As most HIV infection worldwide is to be found in developing countries (*Figure 6.1*), the vaccine would have to be inexpensive and a minimum number of immunizations necessary.

Several strategies are currently being explored. By far the greater number of experiments have been performed using killed whole virus, modeled on the Salk poliomyelitis vaccine. The virus is inactivated with a number of different agents including formalin, Tween-ether (Tween is a non-ionic detergent), β-propriolactone and UV light. Active immunization is greatly facilitated when the inoculation is presented in association with an adjuvant. Various formulations have been tested, for example, muramyl dipeptides, incomplete Freund's adjuvant, immune stimulating complex, and alum (which is the only one of these agents currently licensed for human use). Killed SIV, either from native isolates or molecularly cloned preparations, injected into macaques gives some protection against low doses of the same strain but the best results were obtained if the challenge was not given until 12 months after the final immunization. Other studies demonstrate that killed virus affords protection against intravenously injected virus but not to virus delivered to mucosal surfaces. This will require further investigation. The use of conventionally attenuated live SIV does not prevent infection against a virulent strain but does prevent disease. More encouraging results were obtained when rhesus monkeys were immunized with a live SIV clone containing a *nef* deletion. These animals were challenged with a pathogenic SIV strain 2 years after the final inoculation and have remained healthy since that time, with normal CD4 counts, for more than 4 years. It is unlikely that a vaccine based on a live

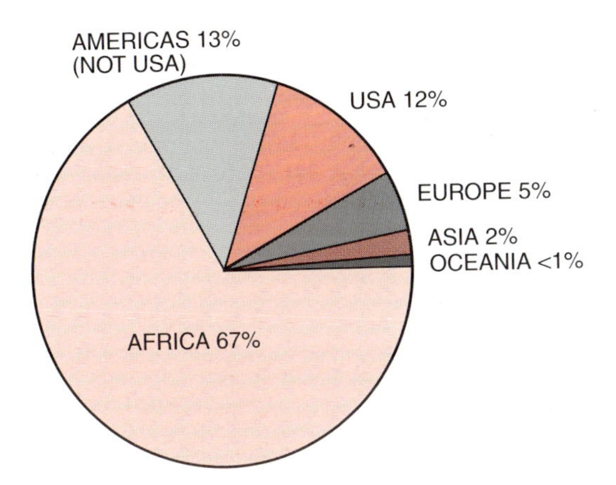

Figure 6.1: Geographical distribution of AIDS cases. Adapted from WHO AIDS data at 31 October 1993.

attenuated version of HIV-1 would be approved for use in humans until all methods employing non-infectious agents have been exhausted.

Experiments with purified HIV envelope glycoprotein, expressed in a baculovirus or Chinese hamster ovary cell line and having near normal glycosylation, have shown limited success. Although it proved to be safe as an immunogen, antibody titers to gp160, following multiple immunizations in seronegative individuals, were low and diminished rapidly over a 2-year follow-up period. Vaccinia virus has been used as a vector, where gp160 is produced in host cells in association with vaccinia virus infection. This has proceeded as far as a phase 1 clinical trial, inducing both cellular and humoral immunity, and has a more vigorous effect in individuals who had not been previously exposed to vaccinia, for instance, through smallpox immunization. Those individuals with prior exposure could not be boosted with a second live vaccinia–gp160 immunization, as the response to the primary challenge was so strong. However, the use of recombinant gp160 as a boosting agent proved more effective against challenge with SIV when the same experiment was repeated in macaques using vaccinia–SIV gp160 as the primary immunizing agent. Using recombinant gp120 as immunogen showed no more promise than gp160. While the immunogen was well tolerated, disappointingly low titers of non-neutralizing antibodies were produced, although the cell-mediated response to gp120 was fairly strong. In experiments where chimpanzees were immunized with human-derived recombinant gp120 and the challenge was with the homologous HIV isolate, protective immunity was obtained if the neutralizing antibodies were directed at the V3 loop of gp120.

Most current clinical trials employ recombinant subunits of HIV-1 or HIV-2. Most of these recombinant vaccines have failed to elicit protective immunity in the SIV model in contrast to the apparent success of HIV-1 *env* subunit immunogens in the chimpanzee. The structural, and therefore immunogenic, differences between SIV and HIV envelope glycoproteins may account for these discrepancies. The core protein p24 has been used in vaccine trials and has provided some encouraging results. Cynomolgus macaques immunized with HIV p24 preparation possess CD4+ T-cells that recognize both pure p24 and whole inactivated HIV-1. Co-infection of cells with vaccinia–HIV p24 or vaccinia–gp160 constructs induces the production of virus-like particles that express gp160 but lack the HIV genome, and have the potential to induce suitable immune responses without the danger of viral infection.

Unique determinants are to be found within any given Ig molecule. They are found in all isotypes and are known as idiotypes. Specific antibodies to these epitopes are called anti-idiotypes and are thought, in some cases, to contribute to the pathogenic effects of HIV infection. One autoimmune anti-idiotype-like antibody directed against an epitope on anti-gp120 antibody has been implicated in thrombocytopenia. This principle has been used in vaccine trials. Monkeys were immunized with the anti-idiotype antibody described above in order to induce antibodies that would bind to the CD4 binding site on HIV-1 gp120. The antibody so produced neutralized HIV activity *in vitro* and further clinical trials in humans are ongoing.

It is clear from vaccine trials to date that the development of a successful AIDS vaccine is particularly difficult (*Figure 6.2*). Diversity of the virus is one of the major problems: there are at present five known main families of HIV-1 and the difference is as great as 30% in the *gag* and *env* gene sequences between the families. Three families are to be found across the African continent, one in Zambia and Somalia, one in Uganda, Ivory Coast and Kenya, and one in Zaire. The virus family in Zaire has also been found in Brazil. The USA and Europe share a family of which the MN isolate is the most common, and is also found in some parts of Africa. Taiwan has its own unique family. A vaccine based on isolates of a European strain of HIV-1 may not be effective when used in other areas of the world. It may be possible to isolate a virus that has a common epitope that will crossreact with other isolates from across the world; peptide data banks are constantly being updated and scanned to this end.

The affinity of viral gp120 for the host-cell CD4 molecule is so high that if a vaccine is given after exposure, as is the case with rabies, it is unlikely to prevent infection, but if it is given before and after infection it may assist the immune system in slowing the progression of disease. Certainly, research aimed at the production of a cheap, effective vaccine that gives protection against the greatest number of isolates worldwide should be high on the list of priorities. In the meantime, education about the modes of transmission

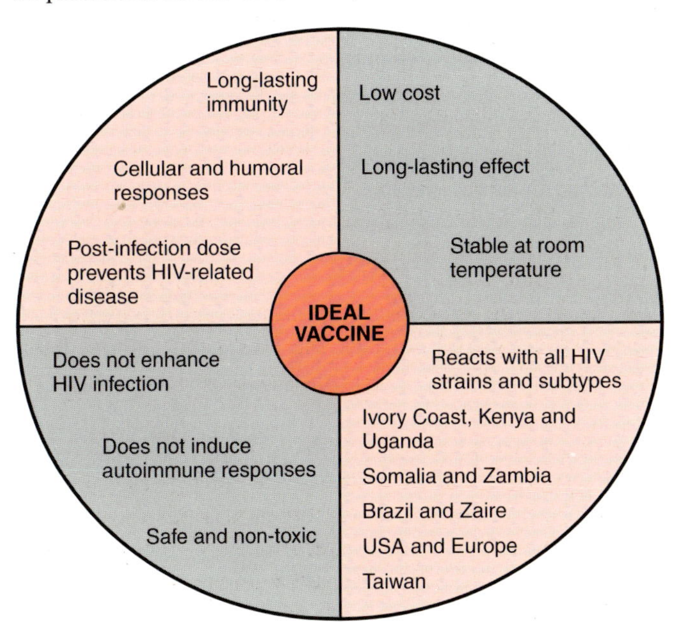

Figure 6.2: The ideal HIV vaccine.

of the virus, accessibility to barrier methods of protection during sexual intercourse, and the development of topical HIV-cidal gels must continue.

6.6 The special case – perinatal transmission

Almost 2000 infants per year are born infected with HIV in the USA alone. Worldwide, the number of children infected with HIV-1 since the syndrome was first described exceeds one million. Studies on the incidence of transmission from mother to offspring present figures which range between 10 and 50% (see Chapter 1). Whether these infections occur *in utero* or at birth, vaccination of the mother during pregnancy may protect the infant. For over a decade perinatally transmitted hepatitis B has been prevented by the combined use of hepatitis B virus Ig and a vaccine based on the hepatitis B virus. More than 90% of infants born to hepatitis-B-positive mothers are protected by this approach. In another study it was found that pregnant women vaccinated against *Hemophilus influenzae* type b (Hib) during the third trimester of pregnancy gave birth to Hib-immune offspring. This was due to high levels of maternal anti-Hib crossing the placenta. The greatest concentrations of antibody cross the placenta after the thirty-second week of pregnancy. If a safe, effective vaccine against HIV-1 were to be employed in a similar regime, the risk of infection through exposure to maternal blood at birth would be minimized. Before such a strategy of immunoprophylaxis could be established it would first be necessary to determine whether infection occurs *in utero* or at birth so that immunization could be given at the appropriate time.

Other factors may also affect the transmission of infection from mother to child. As discussed previously (Chapter 2), mothers with a high titer of high-affinity antibody to the principal neutralizing determinant of the V3 loop of gp120 are less likely to transmit HIV-1 vertically to their offspring, an observation that may assist in the design of an effective prophylactic vaccine. Clinical trials are currently proceeding under the direction of the National Institutes of Health (NIH) AIDS Clinical Trial Group, in which anti-HIV Ig is given to the mother during pregnancy and to the infant within 12 hours of delivery in conjunction with the anti-retroviral agent AZT, the mother receiving a dose during delivery and the baby during the first 6 weeks of life. The Ig to be employed in the immunization program will be checked using PCR. Drawing on experience gained by using passive immunization for hepatitis B, allied to stringent safety precautions, the use of passive immunization with HIV Ig in pregnant women and their offspring may prove to be a breakthrough in one aspect of HIV prophylaxis.

Reference

1. Parkin, J., Underhill, J., Eales, L-J. and Pinching, A.J. (1992) in *VIII International Conference on AIDS/III STD Conference, Amsterdam* (Abstract).

Appendix A. Glossary

Adjuvants: substances which enhance the immune response to immunogenic antigens. They may increase the surface area of the antigen, as is the case with alum-precipitated diphtheria toxin or they may prolong the retention time within the body by causing a depot effect, as with Freund's adjuvant. Certain adjuvants may even accentuate one arm of the immune response preferentially.

Allogenic: genetic variation within a given species.

Antibody: immunoglobulin molecule synthesized by B-lymphocytes which binds specifically to its antigen. Five main isotypes exist in man, IgG (with four subclasses, IgG1, IgG2, IgG3 and IgG4), IgA (with two subclasses, IgA1 and IgA2), IgM, IgD and IgE.

Antigen: any molecule that can stimulate the production of a specific immune response.

Antigen-presenting cell (APC): a number of different leukocytes are able to present antigen in a form that can stimulate lymphocytes. These include Langerhans cells in the skin, follicular dendritic cells which present antigen to B-cells in the germinal centres of lymph nodes, and macrophages which can act as APCs in the outer cortex of the lymph node.

Antiretrovirals: any agent that prevents replication of a retrovirus. Reverse transcriptase is unique to retroviruses and is therefore a target in antiviral therapy. In 1985 it was shown that several members of a family of compounds, the dideoxynucleosides, were able to inhibit HIV replication. Unlike physiological deoxynucleosides these compounds have an azido (N_3) group at the normal 3′ hydroxyl (OH) group which does not allow phosphodiester linkage, thus inhibiting viral transcription through its reverse transcriptase. Dideoxynucleosides used to date include: 3′-azido-2′,3′-dideoxythymidine (AZT or zidovudine), 2′,3′-dideoxyinosine (ddI), 2′,3′-dideoxycytidine (ddC) and 2′,3′-didehydro-2′,3′-dideoxythymidine (D4T).

Apoptosis: see Programmed cell death.

Asymptomatic: lacking in clinical symptoms or signs of disease.

Atopy: clinical conditions associated with type 1 hypersensitivity reaction (IgE-mediated), mainly, asthma, eczema and allergic rhinitis.

Cell-mediated immunity (CMI): lymphocyte or macrophage-mediated reaction to intracellular pathogens where antibody plays a small part only. A number of cytotoxic T-cells are antibody independent, as well as natural killer (NK) cells and lymphokine activated killer (LAK) cells. Other cytotoxic cells are antibody dependent (ADCC) and perform their killer function on antibody-coated target cells.

Clathrin-coated pits: clathrin is a non-glycosylated protein which forms a polyhedral framework on the outer surface of the vesicles that carry proteins between the endoplasmic reticulum, the Golgi apparatus and their final destination. The clathrin cage can be reversibly dissociated into three-legged complexes known as triskelions. The influenza virus, which has been studied extensively, ribosomally synthesizes its viral spike glycoproteins on the endoplasmic reticulum. After further processing in the Golgi apparatus these proteins (HA and NA) are transported in clathrin-coated vesicles to specific areas of the plasma membrane. Here they bind with a nucleocapsid binding shell and bud from the cell surface to form the mature virion.

Clusters of differentiation (CD): antigens found on the surface of animal cells, first identified with monoclonal antibodies. CD antigens are validated, once they have been cloned and sequenced at the cDNA level, at international workshops. At the present time there are more than 100 such clusters. Those particularly significant in HIV infection include CD3 which associates with the T-cell receptor (TCR) and recognizes peptide antigen bound to MHC antigens. The MHC class II antigens, together with CD4 antigen, transduce the signals necessary for T-cell activation. CD8 is expressed on approximately one-third of mature T-lymphocytes and acts as a co-receptor with MHC Class I restricted TCRs in antigen recognition and T-cell activation. The CD28 antigen appears to provide an essential co-stimulatory signal to the CD3/TCR signal. Without its signal, TCR stimulation leads to anergy or apoptosis.

Cytokines: a family of low-molecular weight (mostly <25 kDa) proteins, usually glycosylated, that regulate most biologically important processes. They are related through their function rather than structurally. Examples include the interleukins, tumor necrosis factor and the interferons.

Endocytosis: is the process by which phagocytic cells internalize microorganisms following engulfment and encapsulation in a phagosome.

Epitopes: that portion of an antigen that is bound to a paratope of an antibody. A single determinant of which any antigen may have many.

Epstein–Barr virus (EBV): a transforming virus that is able to immortalize B-lymphocytes and is the causative agent of Burkitt's lymphoma and infectious mononucleosis. The B-cell carries a surface determinant known as CR2 (CD21), a receptor for the complement component C3b to which EBV is also able to bind and thus gain entry to the cell.

Genome: the entire genetic information of a cell.

Genotype: the genetic information inherited from both parents not all of which is necessarily expressed in the offspring.

Glycoprotein: covalent association of carbohydrates with a protein. While the protein is synthesized under genetic control the carbohydrates are added post-translationally and are therefore heterogeneous. The major glycoproteins associated with HIV are gp41 and gp120 which are derived from a precursor glycoprotein gp160. The gp120 forms the external surface envelope protein and the gp41 the transmembrane protein.

GM–CSF: granulocyte–macrophage colony-stimulating factor, a member of the family of colony-stimulating factors that regulate hemopoiesis and includes macrophage colony-stimulating factor (M-CSF), granulocyte colony stimulating factor (G-CSF) and erythropoietin or red cells. These same factors act on mature cells to cause activation.

Herpesviruses: a family of DNA viruses, the main ones in man being, herpes viruses 1 and 2 (which cause cold sores and genital lesions), cytomegalovirus, varicella zoster (which causes chicken pox when a primary infection, and shingles as a reactivation disease), EBV (which causes glandular fever, and is also associated with naso-pharyngeal carcinoma and Burkitt's and non-Hodgkin's lymphoma) and human herpesvirus 6 (which infects CD4+ lymphocytes, but is of uncertain pathogenicity).

Humoral immunity: in the higher vertebrates the humoral immune system usually refers to that part of the process that is largely antibody mediated. The B-cells produce five antibody isotypes, IgM, IgD, IgG, IgA and IgE. There are four subclasses of IgG and two subclasses of IgA. It is presumed that the diversity evolved to keep pace with the anatomical complexity of the organism and the ever increasing load of environmental antigens. Invertebrates appear to lack immunoglobulins but have other humoral factors such as agglutinins and lysins. MALT, the mucosa-associated lymphoid tissue can also be regarded as part of the humoral immune system.

Idiotype: is the antigenic property of the variable region of an antibody. Idiotypes are usually specific to a B-cell clone and determine the specificity of the antigen binding site.

Interferons: are glycoproteins secreted by virally infected cells. The secreted interferon then binds to specific receptors on the surface of other cells, transducing a signal which causes impairment of viral replication to both RNA and DNA viruses. The interferons, α, β and δ are produced by leukocytes and fibroblasts and interferon γ is made only by lymphocytes.

Interleukins: once described as proteins produced by leukocytes that act upon other leukocytes but subsequently shown to be produced by and to act upon non-leukocytes. They are all designated IL-*n*, where *n* is a number assigned at an international workshop.

ISCOMS: **I**mmune **S**timulating **COM**plexe**S** are cage-like structures formed by mixing HIV gp120 envelope (or other) protein with purified saponin. The saponins are a family of heavily glycosylated triterpenes that behave like detergents at the cell surface. They are usually derived from the plant *Quillaia saponaria*. ISCOMS act as an adjuvant and, when highly purified saponin is employed together with alum and gp160, enhanced cell-mediated responses have been observed.

Lymph nodes: act as a filter, clearing antigen from the interstitial tissue fluid and lymph on its journey from the periphery to the thoracic duct. Blood vessels enter and leave at the indentation or hilus together with several afferent lymphatic vessels and a single efferent lymphatic vessel. The node is encapsulated with collagen and immediately below the capsule is a sinus lined with phagocytic cells. Antigen-bearing lymphocytes enter the node via the afferent vessels and pass through this sinus. The cortex contains B-cells in clumps which form the primary follicles and those clumps with an active proliferation site or germinal centers are the secondary follicles.

Lymphocytes: are white blood cells responsible for specific recognition of pathogens. They all emanate from bone marrow stem cells, the T-lymphocytes develop in the thymus and, in mammals, the B-lymphocytes develop in the bone marrow. B-cells are able to divide and differentiate into plasma cells which produce antibodies while T-cells are pluripotent. Some react with B-cells and help them in their antibody-producing role, others assist monocytes in their phagocytic duty. These are helper cells. Another group of T-lymphocytes are able to destroy infected host cells and these are known as cytotoxic T-cells.

Macrophage: is a member of the mononuclear phagocytic cell line. Like the lymphocyte it is derived from the bone marrow stem cell and has the ability to engulf, internalize and destroy infectious agents. It is also able to present antigen to T-cells. When in the blood, the cell derived from the bone marrow is called a monocyte. This cell is able to migrate to the tissues where it is termed a tissue macrophage or when it is a specialized cell it is given other names such as Kupffer cell in the liver.

Major histocompatibility complex (MHC): is the genetic region whose product is responsible for rejection of foreign tissue and functions as a signalling molecule between lymphocytes and antigen-presenting cells.

Monocyte: see macrophage.

Natural killer (NK) cell: a specialized lymphocyte involved in the non-antigen-specific killing of tumor cells (tumor surveillance) and virally infected cells.

Neutralizing antibodies: are those antibodies that have the ability to prevent or inhibit viral replication *in vitro*.

Non-Hodgkin's lymphoma: lymphoma which is usually of B-lymphocyte lineage and often expresses EBV proteins. It shows a marked increase in incidence in the immunosuppressed.

Nuclear regulatory factors: or response elements bind to specific sites in DNA to regulate gene transcription. Nuclear factor of activated T-cells (NF-AT) is not expressed constitutively but is induced following TCR-CD3 triggering. Another nuclear factor which plays a critical role in T-cell activation, NF-κB, is inducible in T-cells and is negatively regulated by an inhibitor protein, I-κB. The enhancer for NF-κB is found in both the IL-2 receptor promoter and the HIV-1 long terminal repeat and their similar NF-κB-mediated inducibility in T-cells by mitogens forms a basis for the observation that T-cell activation is accompanied by enhanced HIV-1 replication.

Oncogene: a gene coding for a protein involved in cellular growth control, which, if over- or under-expressed, can contribute to transformation. The normal cellular equivalent is termed a **proto-oncogene**.

Oncovirus: is able to transform normal cells into malignant cells. The best studied of these viruses is the Rous sarcoma virus which is a retrovirus. The chromosome contains four genes, three of which are essential for viral replication. The fourth gene, v-*src* (viral sarcoma) encodes a 60 kDa protein, $p60^{v\text{-}src}$ which mediates host cell transformation and is therefore termed an oncogene.

Opportunist infections: are *either* infections caused by organisms that are usually non-pathogenic (for example the severe pneumonia that is caused by *Pneumocystis carinii* in AIDS patients or those with leukemia) *or* atypically severe infections caused by organisms can cause disease in immunocompetent individuals (for example, cytomegalovirus which can cause a mild glandular fever-like illness in anyone, but leads to life-threatening illnesses in the immunocompromised). Opportunist tumors refers to the tumors observed in immunosuppressed individuals, which often are suspected of having a viral aetiology.

Phenotype: is the appearance or character of an individual resulting from those parts of the genotype actually expressed.

Polymerase chain reaction (PCR): is a technique by which a specific segment of DNA of up to 6 kb can be amplified. Briefly, a denatured DNA sample is incubated with a DNA polymerase in the presence of two oligonucleotide primers to direct the polymerase to synthesize new complementary strands. This process is repeated many times leading to exponential amplification of the starting copy. Twenty-five cycles of PCR amplification increases the amount of target sequence by about a millionfold. Between each cycle the two strands of the duplex DNA are separated by heat denaturation.

Programmed cell death (apoptosis): is a mechanism of cell suicide which occurs as part of many developmental and immunoregulatory processes. It requires cell activation, maybe through one receptor alone, *de novo* protein synthesis and activation of an endogenous endonuclease which cleaves nuclear DNA into 200 bp fragments. At this stage the cell fragments are readily phagocytosed.

Prophylaxis: is preventative treatment against disease.

Protein kinases: transfer phosphate groups to the amino acids serine, threonine and tyrosine and thereby change the nature of the protein to which these amino acids belong. This may lead to activation or inhibition of the protein concerned, or may make binding to a target protein more favorable.

Resistance: in antimicrobial therapy, this is the ability of an organism to overcome the actions of antimicrobial drugs, by changes in the targets at which the drug is directed. This may occur by mutations in the genetic material of the organism, or by transfer of genetic resistance from organism to organism by plasmids. Multidrug resistance refers to the lack of sensitivity in *in vitro* testing to more than one antimicrobial agent.

Retroviruses: are RNA-containing eukaryotic viruses, many of them being tumorigenic, that contain an RNA-directed DNA polymerase known as reverse transcriptase (see below).

Reverse transcriptase: was discovered independently by David Baltimore and Howard Temin in 1970. It synthesizes DNA in the 5′ to 3′ direction from primed templates, RNA being the template in the case of reverse transcriptase.

Sense and antisense: it is usual for RNA synthesis to be initiated only at specific sites on the DNA template. This was shown in early experiments where DNA from the bacteriophage $\phi\chi174$ was hybridized with RNA produced by $\phi\chi174$-infected *E. coli*. This bacteriophage carries a single 'plus' strand of DNA which, in *E. coli*, directs the synthesis of a complementary 'minus' strand which combines with the plus strand to form a circular duplex DNA, known as the replicative form. The RNA produced by $\phi\chi174$-infected *E. coli* does not hybridize with DNA from the intact phage but will hybridize with the replicative form. Thus the minus strand acts as a template and is transcribed while the plus strand is not transcribed. The minus strand is therefore called the **sense** strand and the plus strand the **antisense** strand.

Seroconversion: the development of serum antibodies to a specific organism.

Syncytia: the fusion of more than one cell to another to form a multinucleated giant cell.

T-cell receptor (TCR): consists of either α/β or γ/δ heterodimers which are clonotypic and have immunoglobulin-like constant and variable regions. These associate with invariant membrane-spanning ζ/ζ or ζ/η chains. The α/β and γ/δ dimers recognize antigen bound to MHC antigens and signals transduced subsequently through the invariant ζ chains lead to T-cell activation.

Th_1 and Th_2: subsets of helper T-lymphocytes characterized by the cytokine secretion pattern and mainly CD4 positive. Lymphocytes that secrete mainly IL-2 and IFN-γ on stimulation are classified as Th_1, those that secrete IL-4 and IL-5, as Th_2, and those that have a mixed pattern as TH_0.

Tumor necrosis factors (TNF): cytokines, secreted by many cell types, that were initially identified by their ability to cause tumor necrosis. TNFα is also known as cachetin, and TNFβ, as lymphocytotoxin.

Virus: an intact virus particle is known as a virion and is composed of a nucleic acid molecule enclosed in a protein case or capsid. In complex virions this capsid is surrounded by a lipid bilayer and a glycoprotein-containing envelope. In contrast to the host cell, in which the hereditary molecules are invariably double-stranded DNA, the nucleic acid of the virus may be either single- or double-stranded DNA or RNA.

Index

ALSO AVAILABLE FROM BIOS SCIENTIFIC PUBLISHERS LTD

Autoimmunity

W. Ollier & D.P.M. Symmons
University of Manchester, Manchester, UK

Every clinician needs some familiarity with autoimmunity. This book recognizes that need and is an informative guide to autoimmunity for the non-specialist. Starting at the level of the gene, the book progresses to that of the cell, and then goes on to look at the consequences of autoimmunity on the whole organism (both animal and human). The book concludes by considering the long term prognosis and exploring the treatment options.

"... intelligent, well-written, clearly illustrated and enjoyable" *Br.J.Biomedical Science*

Contents

The immune system; Genetic basis of autoimmune disease; Nature of auto-antigens; Consequences of autoimmunity at cellular and humoral levels; Multi-system disease; Disease of predominantly one organ; Long term prognosis and treatment; Further reading; Glossary.

Of interest to:

Medical students; Postgraduates and non-specialist researchers in medicine.

Paperback; 152 pages; 1-872748-50-3; 1992

ALSO AVAILABLE FROM BIOS SCIENTIFIC PUBLISHERS LTD

Molecular Virology

D.R. Harper
Medical College of St Bartholomew's Hospital, London, UK

Molecular virology is a rapidly moving field and recent developments in understanding the basic processes of viral infection have provided a vast range of information which can be very confusing to the non-specialist or to anyone new to the area. This book is a much needed introductory text summarizing current knowledge of molecular virology in key areas.

Topics covered include fundamental virology, immunology and pathogenesis, vaccines and antiviral drugs, novel diagnostic techniques, cloning and the use of viral vectors. In addition, the book provides a comprehensive survey of likely future developments, particularly those in vaccines, antiviral drugs and viral delivery systems.

Contents

Virus structure and replication; Medical applications - antiviral strategies; Medical applications - nucleic acids and cloning; future prospects.

Of interest to:

Medical students and junior hospital doctors undertaking specialist training or research in virology, microbiology or infectious diseases; Undergraduates and postgraduates on other courses with virological components; Clinical scientists and medical laboratory scientific officers taking specialist training in virology.

Paperback; 168 pages; 1-872748-57-0; 1994

PCR

C.R. Newton & A. Graham
Zeneca Pharmaceuticals

The Polymerase Chain Reaction (PCR) has revolutionized and underpins modern molcular biology. This up-to-date text provides both an introduction to PCR and authoritative coverage of the wide range of current applications. By focusing on current PCR technology, the book is essential reading for both first-time users and experienced researchers. Given the increasing importance of PCR, it will also rapidly become an invaluable text for undergraduates.

"I regard the book to be a very useful text and reference for new and experienced PCR users, and would rate it well above other recent texts that I have examined on the subject." *Trends in Cell Biology*

Contents

What is PCR?; Instrumentation, reagents and consumables; Amplifying the correct product; Cloning of PCR products; Isolation and construction of DNA clones; Modification of PCR products; Joining overlapping PCR products; PCR mutagenesis; Sequencing PCR products; New sequence determination; Detecting pathogens; Analysis of known mutations; Characterizing unknown mutations; Fingerprinting; Human genome mapping applications; Quantitative PCR. *Appendices:* Suppliers; Glossary; Further reading.

Of interest to:

Undergraduates; postgraduates; and researchers.

Paperback; 176 pages; 1-872748-82-1; 1994

ALSO AVAILABLE FROM BIOS SCIENTIFIC PUBLISHERS LTD

The Human Genome

T. Strachan
St Mary's Hospital, Manchester, UK

If you have any interest in the Human Genome Project, this book is a must! A clear introduction to the structure of the human genome and the ways in which recent knowledge is influencing medical research and practice.

"...an easy to read, up to date and complete treatment of the human genome from biochemistry to medical genetics." *Endocrinologist* "...a crash course on the ways in which classical human genetics is being revised and extended by the methods of applied molecular biology" *Biologist*

Contents

Organization and expression of the human genome; Evolution and polymorphism of the human genome; Analyzing human DNA; Mapping the human genome; Human disease genes - isolation and molecular pathology; The human genome - clinical and research applications.

Of interest to:

Undergraduates; Postgraduates; Researchers; Medical students.

Paperback; 170 pages; 1-872748-80-5; 1992

ORDERING DETAILS

Main address for orders

BIOS Scientific Publishers Ltd
St Thomas House, Becket Street,
Oxford OX1 1SJ, UK
Tel: +44 1865 726286
Fax: +44 1865 246823

Australia and New Zealand
DA Information Services
648 Whitehorse Road, Mitcham, Victoria 3132, Australia
Tel: (03) 873 4411
Fax: (03) 873 5679

India
Viva Books Private Ltd
4346/4C Ansari Road, New Delhi 110 002, India
Tel: 11 3283121
Fax: 11 3267224

Singapore and South East Asia
(Brunei, Hong Kong, Indonesia, Korea, Malaysia, the Philippines, Singapore, Taiwan, and Thailand)
Toppan Company (S) PTE Ltd
38 Liu Fang Road, Jurong, Singapore 2262
Tel: (265) 6666
Fax: (261) 7875

USA and Canada
Books International Inc
PO Box 605, Herndon, VA 22070, USA
Tel: (703) 435 7064
Fax: (703) 689 0660

Payment can be made by cheque or credit card (Visa/Mastercard, quoting number and expiry date). Alternatively, a *pro forma* invoice can be sent.

Prepaid orders must include £2.50/US$5.00 to cover postage and packing for one item and £1.25/US$2.50 for each additional item.